T0359114

RAMMED EARTH CONSTRUCTION

PROCEEDINGS OF THE FIRST INTERNATIONAL CONFERENCE ON RAMMED EARTH CONSTRUCTION, PERTH AND MARGARET RIVER, WESTERN AUSTRALIA, 10–13 FEBRUARY 2015

Rammed Earth Construction

Cutting-Edge Research on Traditional and Modern Rammed Earth

Editors

D. Ciancio & C. Beckett
University of Western Australia, Australia

CRC Press
Taylor & Francis Group
Boca Raton London New York Leiden

CRC Press is an imprint of the
Taylor & Francis Group, an **informa** business

A BALKEMA BOOK

CRC Press/Balkema is an imprint of the Taylor & Francis Group, an informa business

© 2015 Taylor & Francis Group, London, UK

'Rammed Earth in a concrete world' and 'Rammed earth thermodynamics' by M. Krayenhoff
© 2015 Sirewall, Salt Spring Island, BC, Canada. Used with permission

Typeset by V Publishing Solutions Pvt Ltd., Chennai, India
Printed and bound in Great Britain by CPI Group (UK) Ltd, Croydon, CR0 4YY

Published by: CRC Press/Balkema
P.O. Box 11320, 2301 EH Leiden, The Netherlands
e-mail: Pub.NL@taylorandfrancis.com
www.crcpress.com – www.taylorandfrancis.com

ISBN: 978-1-138-02770-1 (Hbk)
ISBN: 978-1-315-69294-4 (eBook PDF)

Table of contents

Preface

The First International Conference on Rammed Earth Construction (ICREC 2015) is organized by the University of Western Australia (UWA). It is a four-day event that consists of a 2-day workshop held at the UWA Trinity College in Perth, and a 2-day conference held in Margaret River, Western Australia.

The ICREC 2015 workshop aims to promote the use of rammed earth in Australia and around the world. Experienced and well-respected rammed earth builders, engineers and architects have been selected as invited speakers to share their valuable knowledge on a variety of topics including: construction methods; standards and guidelines; laboratory procedures; heritage and conservation and thermal performances, analysis and design. The first two keynote papers contained in these proceedings from S. Dobson and D. Easton are part of the workshop event.

The ICREC 2015 technical conference is host in Margaret River, Western Australia, a region that has the largest concentration of rammed earth buildings and structures. The conference brings together academics and practitioners in order to communicate the latest developments in the design and analysis of rammed earth structures. These proceedings contain the keynote papers by Dr. Augarde and Dr. Hu, and the works presented in the technical conference in Margaret River. Each paper has been independently and anonymously reviewed by to 2 peers from within industry and academia who are experts in the relevant fields.

The editors would like to thank all the keynote speakers and authors for their effort and support for this conference. The editors are grateful to Ms. Dawn Feddersen for her assistance with secretarial duties for the ICREC 2015 event.

Acknowledgements

A successful conference requires the involvement of a large number of people. The editors would like to thank the following individuals, companies and groups for their support in making this conference possible.

Sponsorship has gratefully been provided by:

UWA Faculty of Engineering, Computing and Mathematics
Pritchard and Francis
Earth Building Association of Australia

Local Organising Committee (LOC):

Daniela Ciancio, *Conference Chair, University of Western Australia, Australia*
Chris Beckett, *Conference Vice-Chair, University of Western Australia, Australia*
Dawn Feddersen, *Conference Secretary, University of Western Australia, Australia*
Kenneth Kavanagh, *Formerly at University of Western Australia, Australia*
J. Antonio H. Carraro, *University of Western Australia, Australia*
William Smalley, *Scott Smalley Partnership, Australia*

Papers were reviewed by the LOC and by members of the International Advisory Board:

Charles Augarde, *Durham University, UK*
Gabriel Barbeta Sola, *Universitat di Girona, Spain*
Alan Brooks, *Perth Stabilised Earth, Australia*
David Easton, *Rammed Earth Works, USA*
Domenico Gallipoli, *Université De Pau Et Des Pays De L'adour, France*
Matthew Hall, *University of Nottingham, UK*
Andrew Heath, *University of Bath, UK*
Peter Hickson, *Earth Building Solutions, Australia*
Paul Jaquin, *LDE Consulting, New Zealand*
Chintha Jayasinghe, *University of Moratuwa, Sri Lanka*
Meror Krayenhoff, *SIREWALL, Canada*
Ying Lei, *Xiamen University, China*
Claudio Mazzotti, *University of Bologna, Italy*
Jean Claude Morel, *Université de Lyon, France*
Hugh Morris, *University of Auckland, New Zealand*
David Toll, *Durham University, UK*
B.V. Venkatarama Reddy, *Indian Institute of Science, India*
Peter Walker, *University of Bath, UK*
Bly Windstorm, *Earth Dwell Ltd., USA*

Keynote papers

Rammed Earth Construction – Ciancio & Beckett (Eds)
© 2015 Taylor & Francis Group, London, ISBN 978-1-138-02770-1

Rammed earth in the modern world

S. Dobson
Director, Ramtec Pty. Ltd., Australia
Vice President, EBAA, Australia

ABSTRACT: Modern rammed earth has a long and successful history. Rammed earth is arguably the most popular form of building that exists on the planet but in some modern countries it is presently considered alternative or new and innovative. Modern engineering is being applied to structural design, thermal modelling, mix design, construction methodology and other aspects of an ancient building method that is now truly revived. Modern architecture has captured the essence of the past and the present to produce a bewildering array of buildings spread across the globe of amazing complexity and stunning beauty. The author has since 1976 built over 750 rammed earth structures in Australia and describes some old and some new buildings from around the world whilst outlining the many benefits of rammed earth, some sticking points that need resolution and some future predictions of where this expanding new industry can go.

1 INTRODUCTION

Leonard Cohen sang "First we take Manhattan, then we take Berlin".

Taking Manhattan:
The unfired earth building industry built the first Manhatten in the desert of Yemen at Shibam, over 1500 years ago with local earthen materials, all unstabilised and using the procedures of mud brick, cob and rammed earth (Figure 1). Over 500 buildings to 14 stories high. Some walls 1.2 m thick. All UNESCO World Heritage Buildings at Wadi Hadramaut, mostly still occupied, all walls fully loadbearing, held together with just clay and in an earthquake area (Earth Architecture 2014).

Taking Berlin:
14 years ago unstabilised rammed earth took Berlin with the Chapel of Reconciliation, the first modern rammed earth building in Germany for 50 years. Rammed earth contractor Martin Rauch, backed by a group of Engineers, Architects and others built the curved loadbearing unstabilised rammed earth walls and floor of the Berlin Chapel of Reconciliation. This was considered a major breakthrough in Germany, a country with thousands of old successful rammed earth buildings including House Rath in Weilburg, Germany, 7 stores high, loadbearing and built in 1828, which is still in use.

Rammed earth taking Manhattan, then Berlin, then the world:
Rammed earth exists historically in nearly every country in the world and is now actively being built new many countries worldwide. There are 7 billion people in the world and it is generally said that one third to one half live in earth buildings (earth buildings meaning: mud brick, rammed earth and cob plus about 20 other techniques), say 3 billion. South American researchers claim more rammed earth than mud brick worldwide which makes rammed earth the **most popular** building material on the planet. Interestingly it is still listed as an "alternative" building material in most modern countries.

Australia leads the modern world in quality and volume of modern rammed earth, all cement stabilised and almost all load bearing and often unprotected from the elements and over all Australian climatic zones, from the deserts to the snowfields and everything in between. Europe leads the world in modern unstabilised rammed earth, most loadbearing and much of it unprotected from the elements. Stabilisers

Figure 1. Manhatten of the desert: the world's first high rise.

Figure 2. Taking Berlin: the rammed earth Chapel of Reconciliation.

Figure 3. Berlin Chapel: unstabilised loadbearing rammed earth.

other than cement, such as lime, abound but will be given little mention in this discussion as most rammed earth in Europe and in third world countries uses clay as the binder (termed unstabilised) and in Australia cement is the predominant stabiliser by far.

An old Welsh saying for earth buildings was: "Give'er a good hat and stout boots and she'll last forever". This of course referred to unstabilised earth buildings needing a good roof with adequate overhangs and good foundations and details to keep the base of the walls dry, in order to achieve a long life which was generally measured in centuries, for mud brick and for rammed earth. This saying is now only partially true for rammed earth because with modern stabilisation methods (especially cement) and also with unstabilised methods, particularly from Europe, these rules can be relaxed significantly, sometimes totally.

Enormous advances have been made, in recent years so that now rammed earth can be built fully loadbearing in very tall buildings (4 stories load-bearing and more), in buildings of every size and shape, in extreme earthquake areas (including all of New Zealand, some middle eastern countries and North America including San Andreas Fault locations, Los Angeles, San Francisco, Vancouver and elsewhere along the west coast of North America, and including stabilised insulated reinforced rammed earth in most of these regions), in cyclonic wind areas (several Ramtec buildings survived undamaged in a real Category 5 Cyclone Beaufort Scale 12 Hurricane above 280 km/hr in the worst cyclone areas of Australia, fully exposed coastal), achieving very favourable thermal ratings, delivering excellent thermal comfort, totally waterproof, fireproof, very long lasting, extremely beautiful, low carbon, built at high speed of construction and at reasonable cost.

Modern rammed earth is becoming more globalised as Conferences spread the word and allow the forging of contacts and the exchange of information. The internet has played a large part in extending the reach of modern rammed earth from the many modern centres of excellence that include Europe, North America, Australia and elsewhere, to every corner of the globe. Historical methods of building in rammed earth are still valid and can still sometimes give walls that are beautiful, durable, load bearing, and which last for centuries. Such ancient techniques include taking suitable soil directly from the base of the building, dug out at the right time of year so that it is at optimum moisture content for maximum dry density compaction, and without any additive or mixing, hand ramming it using clay as the only binder, into traditional formwork and incorporating suitable wall protective measures (damp proof base course and top capping, controlled erosion breaks etc). Also examples abound of modern successful rammed earth buildings that are built to last, to almost every style of architecture.

In the past, earth was not widely promoted as it was freely available to all and thus not the provenance of serious business. Previously in many countries, earth was seen as a poor mans product and less desirable than the modern industrialised products that abounded after the great wars. No longer are these scenarios valid with increased enviro-awareness plus rammed earth is now a serious business and becoming more mainstream in a growing number of countries. Modern desirable rammed earth buildings are now making it an aspirational product. Promotion of rammed earth is underway and this will grow as sustainability, healthy home, aesthetics, low carbon and other issues become more important in building choices. Rammed earth is itself becoming a modern industrialised product, as wide ranging research moves it ahead. Carried forward by its many advantages. Rammed earth is now often considered as the product of choice from which to construct appropriate environmentally prestigious buildings.

2 RAMMED EARTH IN ANTIQUITY

The oldest buildings made of rammed earth are at Catahyouk near to Konya in Turkey, around 10,000 years old. They were lived in for about 1700 years. Now it is a famous archeological site with current ongoing excavations. Here, as was often the case with old buildings, the constructors were not respectful of the techniques used and evidence is clear of mud bricks of various sizes, with various mortar configurations, cob and rammed earth. I have seen clearly layered ancient ramming lines. All homes here are load bearing with earthen walls and roofs and floors. They are considered to be unstabilised (and clay is the binder) but Turkish rammed earth expert Prof Bilge Isek has been engaged recently to look to see if vegetable, or plant matter, as a stabiliser was used in the construction of the walls. Catalhyouk (Figure 4) is one of the first villages formed by mankind when nomads first settled and began to farm.

China has a history of rammed earth including parts of the Great Wall of China (built some 2500 years ago) and other Chinese rammed earth buildings to 4500 years old. More modern rammed earth circular Hakka structures and other major buildings in China are mere centuries old. Mexico at Teotihuacan has the 1900 year old Pyramid of the Sun about 70 m tall made of some 2 million tons of rammed earth, faced with stone. Watch towers built from rammed earth by Hannibal about 2000 years ago still exist.

The Romans used rammed earth extensively. Cities of rammed earth were built and the method was widely disseminated; for example, Spain has thousands of centuries old rammed earth buildings including the part rammed earth UNESCO listed Alhambra Palace in Spain, parts of which date to Roman times (Figure 5). Compare these

Figure 5. Rammed Earth Alhambra Palace in Spain.

Figure 6. Rauch House: 3 storey loadbearing unstabilised RE.

Figure 7. Royal Automobile Club, Victoria.

to modern RE structures like the Rauch House, a 3 storey loadbearing unstabilised rammed earth home in Austria using 85% of building fabric construction materials off site (Figure 6), and the Royal Automobile Club, Victoria, the biggest modern loadbearing RE building in Australia (Figure 7). The forms may have changed, but the feel is still the same.

Figure 4. Catalhyouk: 10,000 year old rammed earth homes.

3 MODERN RAMMED EARTH

3.1 *Thermal properties*

Rammed earth is generally the cheapest way to deliver high thermal mass walls. High thermal mass assists greatly in achieving comfort conditions. Both in winter and in summer. In winter they absorb the heat of the day, from well designed passive solar sun direction, and re-radiate it at night. Passive solar buildings can collect free sunshine to provide free heating forever so long as there is thermal mass to store it though the collecting time of the winter day prior to discharging it at night when needed. Lightweight materials cannot do this very well, regardless of R value.

In summer thermal mass walls, floors and also roofs even out the day to night temperature fluctuations. Particularly where there are significant variations between daytime and nighttime temperatures, thermal mass equalises and delays temperature transfer through the walls. By opening buildings at night stored heat can be released, a process known as "night purging", allowing heat to be absorbed again the following day. Lightweight highly insulative materials cannot do this well, regardless of their thermal resistance. Any wall mass outside any insulation layer is effectively quarantined from this desirable summer night activity. For example, in insulated brick veneer, which is currently the single most common form of building in Eastern Australia, the outer brick skin is thermally 'lost', and with it the ability to delay, reduce or even eliminate the costly daytime cooling loads, as rammed earth can. Cooling loads in the heat of the day can be reduced or eliminated in many climates (most climates within Australia) and delayed in all climates with common 300 mm monolithic rammed earth walls. With the hottest summer day air conditioning cooling caused electrical loads now established as the key determinant of all modern first world electrical infrastructure (both electrical generation and distribution), rammed earth has a major part to play. Global warming will exacerbate this problem worldwide, and rammed earth is a key part of the solution. In the USA this was realised and David Easton with a Californian electrical supply company built a thick walled monolithic rammed earth home to showcase the advantages.

Successful rammed earth homes have been built in every climate zone in Australia (yes, every: desert to snow) from Zones 1 to 7 of the Buiding Code of Australia (now called the National Construction Code or NCC) which correspond to northern tropic monsoon, high humidity, various desert categories, Mediterranean Climates and temperate warm to mild to cool moving south to alpine.

Unfortunately most computer thermal simulation programs used to approve buildings from a thermal perspective in Australia do not give adequate recognition to high thermal mass and therefore insulation is often required to be installed within the rammed earth walls particularly in the colder Zone 6 of the NCC. This Zone covers the coast of Victoria, plus almost from Sydney to Adelaide, and the S coast of WA. Fortunately insulated rammed earth is quite buildable and is now in widespread use across Australia. Unfortunately mud brick cannot have insulation inserted into the single skin construction without building 2 skins which doubles the cost and makes it uncompetitive. As a result, the USA (with similar min R value legislation), which led the world in modern adobe (mud brick) construction, has seen it's output plummet. The world's biggest adobe manufacturer, the Hans Sumpf Company of Fresno, California, has ceased production. In Australia the mud brick industry centred in Eltham, near Melbourne, in Climate Zone 6 of the NCC has been decimated by modern thermal laws. Most of the few thousand successful mud brick buildings built prior to thermal regulations in Australia would now be illegal to build. This is despite some of them being certified as true net zero carbon buildings, which is as good as you can get. E-Tool of Perth have done Life Cycle Analyses on some of these buildings to show this (eTool 2014). Many countries including the UK are planning legislation to have new buildings certified as net zero carbon, in the years ahead. Rammed earth (and mud brick), done correctly, can do this now.

Using thermal resistance values as the prime parameter leading to measures for thermal comfort is inaccurate, according to the research from The University of Newcastle (NSW) carried out for Think Brick (Page et al. 2011). The Earth Building Association of Australia agree, with significant such comment on their website. Recent Slovakian climate chamber tests of Australian style cement stabilised rammed earth walls with modelled daily outdoor temp fluctuations on the outside (of the test chamber) showed little temp change on the inside of the sample within the 24 hour period i.e. little temp change inside before reversals of temps began again on the outside (Stone and Bagoňa 2013). With insulation in the middle of the wall there was no internal temp change before reversals of temps began outside (Stone, C. pers comms). This backed up the advice of Rob Freeland of AMCER in Melbourne, who often stated that the energy rating systems for approvals in Australia were flawed because there was little temp variation further into the earth wall past the half way mark before the outside driving temp reversed as night followed day. The current research by the University of Western Australia, and the

Western Australian Department of Housing, into the 2 homes build alongside one another in Kalgoorlie in Western Australia may soon verify this again (Ciancio and Beckett 2013, Beckett et al. 2014). The 2 homes are identical and only differ in one having insulation in the centre of the 300 mm thick cement stabilised rammed earth walls and the other being monolithic (no insulation). Both houses are currently being monitored. Also a similar conventional lightweight house nearby in Kalgoorlie is also being monitored. Some day in the future the humidity/wet bulb side of further monitoring of these homes may take place. Humidity levels are so important for human comfort, that this is an important area of research. Rammed earth walls work so well in this area of balancing humidity (absorbing and releasing it) and most conventional building materials are very bad at this. Within all rammed earth there is a humidity flywheel and a separate thermal flywheel concept. Doing full computer thermal simulations of homes and taking in to account the full hygroscopic properties of all materials is currently near to impossible in Australia. Matthew Hall in UK is researching this (e.g. Hall and Allinson (2009)). Once this technique is mastered then further benefits of building in rammed earth may be able to be better quantified.

3.2 Costs

Perth prices today for one square meter of elevation wall area of a 300 mm thick finished monolithic rammed earth wall, of ideal constructability begin at A$275 (including 10% Goods and Services Tax which is $250 excluding tax). In Australia the density is typically around 2 tonnes per sq m with 1.8 t/m³ being a low figure for a lighter cream coloured limestone mix and 2.3 t/m³ being a high figure for a mix using a heavy base aggregate. Cement contents are typically 5 to 10% of the dry mix and strengths are generally contracted to exceed 2.5 MPa and usually exceed 5 MPa. The required strength for a single story heavy tile roofed building with 300 mm thick rammed earth walls without earthquakes or cyclonic winds is less than 1 MPa. Sirewall in Canada lead the world in exotic rammed earth of very high strength and colours and have achieved mixes up to 47 MPa and typically achieve above 20 MPa with below 10% cement. In the highly seismic areas where they operate they are making rammed earth a direct structural substitute for concrete (generic 20 MPa), but with many advantages. They do this, by very careful aggregate type and grain size selection, to facilitate approvals, to meet stringent earthquake codes whilst having insulation within the walls which is needed in their very cold climates, and to allow engineers to use conventional

concrete calculations. Far south in California, David Easton faces less demands for both insulation and strength. In Europe where cement is frowned upon, due to cement manufacture releasing about 1 t of CO_2 per ton of cement and around 7 billion tons of cement being produced worldwide annually, they favour unstabilised rammed earth where all the binding is done with local clay and no additives. Strengths are generally contracted to exceed 2.4 MPa, they generally achieve 2.7 MPa, and 3 MPa is considered not consistently achievable and not needed anyhow.

3.3 Health and hygroscopic properties

Rammed earth is made as a "humid" mix, it is not liquid, it is not mud. It is a zero slump mix. It is placed and rammed, using kneading compaction. It is not poured and vibrated like concrete. The moisture content at ramming must be the optimum to achieve maximum dry density compaction for that rammer/ramming method. A common moisture content at placement is around 10% by soil mass. Once "dry", unstabilised rammed earth has an equilibrium moisture content of around 6–7% which preserves embedded wood as that moisture level is too low to permit decay/rot (Boltshauser and Rauch 2011). Quite separately rammed earth has a powerful ability to absorb and release moisture and to balance humidity. Minke (2006) states that solid fired clay bricks absorb and release so little water by comparison to unstabilised earth walls that they are inappropriate for balancing the humidity of rooms. But unfired earth is highly appropriate for this. Furthermore phase change materials are coming in as high tech wonders to the building industry and yet rammed earth can outperform many of these products, and often more cheaply as well. Suppliers of phase change products can easily calculate the quantity needed of their chemically advanced systems yet such determinations for the infinitely more sustainable earth wall products have no central promoter/calculator and are thus a more difficult task, in quantifying how much earth will do.

3.4 Other properties

The many desirable properties of well constructed rammed earth include:

- Long life;
- Adequate strength;
- Ability to build very high walls and buildings;
- Robustness/durability/abrasion and wear resistance;
- Good acoustic properties both in terms of reverberated sound and also as a barrier to stop the through transfer of sound;

7

- High thermal mass, and also the ability to introduce insulation within the wall (thereby achieving the attractive natural rammed earth face each side whilst allowing a wide range of R values to be selected);
- The ability to adjust the density by using different aggregates and thereby produce different structural and thermal properties;
- Desirable hygroscopic properties unmatched by most "modern" building products;
- The ability to breathe, meaning to absorb and release air thereby removing particulate matter and often producing a healthier air environment;
- Nil toxicity;
- The ability to produce a desirable humidity within a space that is optimal for human habitation ie rammed earth, by absorbing and releasing humidity as it does so well, balances the humidity and it is balanced to a level that is comfortable being not so low as to be dry and unpleasant (40%) and not so high as to also feel overly humid and unpleasant (70%) (Minke 2006). Additionally higher humidity levels in a building like 70 to 80% foster the growth of fungus spores and mould some of which are very unhealthy and some even deadly. With the powerful humidity equalisation benefits of rammed earth quantified, Minke (2006) reports desirable humidity levels of 45% to 60% in a 5 year test on an unstabilised all rammed earth wall home in 1985 in Germany and further reports that when the owner desired a 5% increase in humidity in his bedroom he achieved it by leaving the bathroom door open. In Canada a study by the British Columbia Institute of Technology showed interior relative humidity in a rammed earth walled home (with insulation in the centre of the wall) just above 50%, confirming Minkes work in Germany and demonstrating ideal comfort conditions for habitation and furthermore a humidity so low that mould growth was impossible. All in a Canadian area where mould in conventional timber homes was and is, a significant ongoing problem, with serious health issues;
- Desirable attenuation of various electromagnetic influences that can harm some humans (Minke 2006);
- Lessening of circadian rhythms being upset by electronic interference and increasing the desirable "earth grounded" human condition by sleeping within massive earth walled spaces (Nicole Bijlsma, speaking at EBAA 2014);
- No cavity is required with monolithic rammed earth and so there is no hidden place for hidden vermin, insects or mould;
- Air Changes and Health: using rammed earth homes in free running mode with often open ventilation, which is possible in all climate zones of Australia except perhaps Alpine, gives major health benefits and potentially major energy bill reduction. The thermal mass of rammed earth allows thermal comfort with reasonable air changes in Australia, produces a healthy home and avoids the unhealthy "sealed box" approach, so favoured by the NCC thermal rating programs;
- Environment: Most people respond favourably to a natural environment and rammed earth can provide this. Rammed earth mimics nature in appearance and in other ways. Biomimicry is alive and well with rammed earth and has been for thousands of years (BG 2014). Rammed earth has been acknowledged as providing a relaxing and calming effect on the occupants and thus has been widely used in churches, schools, hospitals, shops, jails and detention facilities (where inmates stay calmer) and the like;
- End of Building Life: totally recyclable with unstabilised returning to the earth from whence it came and with cement stabilised being crushable and re-useable. There is nil toxic residue;
- Fire: very high fire resistance and recognised as such since 1786 when the famous French architect Francois Cointeraux built a gold medal award winning house "an incombustible house that used unstabilised rammed earth as an inexpensive fireproof construction method that was advantageous against the highly flammable timber homes" (Rael 2009) of the time and he went on to publish extensively and to widely promote rammed earth. Modern rammed earth achieving 4 hour fire test ratings has been used in apartment separation as rated fire walls and in bushfire resistant housing which is of increasing importance in a warming global climate;
- Reinforcement: an easily reinforced material with conventional deformed bars as used in concrete but producing a very low propensity to rust of embedded steel, an ability to "protect" wood by maintaining an equilibrium moisture content so low that timber decay (certain to occur above 20% moisture content) is generally eliminated;
- Versatile: able to be used to build walls footings floors and roofs and lending itself to industrialisation as shown by recent developments around the world and which now are increasing;
- Speed: able to be built quickly, using materials often from nearby and sometimes with very little processing of the raw earth and sometimes with local labour using easily learned techniques, with very little water needed and potentially nil wastage of water or other ingredients;
- Beauty: a very beautiful product reaching deep into human consciousness as a visceral/intuitive beauty;
- and all at reasonable cost.

4 CODES, COMPLIANCE AND STANDARDS

On the question of standards there is a gap in Australian regulations. The NCC has a heading of Earth Building and the words "This page has been intentionally left blank". Previously this space had been occupied by CSIRO "Earth Wall Construction" Bulletin 5 Edition 4 however after nil complaints and nil problems with it from either providers or consumers, CSIRO requested that it be withdrawn from the NCC and it was. Without any consultation with the earth building industry, and to the direct and significant detriment of that industry.

Bulletin 5 now rates, with other publications as being a good book but having no legal standing, alongside EBAA "Building with earth bricks and rammed earth in Australia" (EBAA 2008) and SAI Global publication HBD195 "The Earth building Handbook" (Walker and Standards Australia 2002).

In New Zealand there is a suite of 3 Earth Building Standards operating (e.g. NZS (1998)). They are issued by BRANZ of Wellington New Zealand and covering rammed earth (stabilised and unstabilised) in a very prescriptive engineering way due to the unique problems in NZ of severe earthquakes, high rates of and levels of rainfall, high winds accompanying the rain and a liking for unstabilised earth buildings from many "deep green" consumers.

In USA there is "Standard Guide for Design of Earth Wall Building Systems" (ASTM 2010) which covers rammed earth. New Mexico, a southern USA state famous for adobe buildings, has an earth building standard.

In Germany there are new DIN Standards only in the German language, that relate to mud bricks and earth plaster but which do not cover rammed earth. A rammed earth standard is being worked on and Prof Horst Schroeder, who was very involved, told the writer not to expect a standard to cover rammed earth inside 3 years.

In France there is no rammed earth building standard. However Craterre-Ensag at their International Centre for Earth Construction at Grenoble are researching, teaching, promoting and training in general earth building at an extensive level. The PIRATE Project (an acronym for "Provide Instructions and Resources for Assessment and Training in Earthbuilding") is moving ahead on a large scale. The ongoing work of Craterre, together with other French teaching institutions has significant positive benefits for the field of earth building worldwide. On the French Island of Mayotte in the Indian Ocean due west of Perth, there is an earth building standard.

In Morocco there is an earth building standard, written only in French: "Reglement Parasismique Pour Les Constructions en Terre".

The Standards Association of Zimbabwe have a standard SAZ 724: Code of practice for rammed earth structures (SAZS 2001).

It is the writer's opinion that the best and easiest way to get a standard into play for earth building in Australia is to take EBAA (2008) to the NCC and to update it from an engineering viewpoint to NCC requirements so that it can be direct referenced into the NCC. This is an overdue issue for the earth building industry in Australia.

5 THE FUTURE: BARRIERS TO THE MORE WIDESPREAD USE OF RAMMED EARTH WORLDWIDE

The biggest obstacle to rammed earth in Europe, according to Boltshauser and Rauch (2011) is trust, the trust that is generated not by reading but by seeing touching, feeling and doing (building oneself). This requires more construction of modern RE buildings. There is no such problem in Australia. Rauch, a regular visitor to Africa feels that the solution to the problem there is to teach the advanced newer techniques and to show successful high end modern RE buildings so that it moves from a poor choice to an aspirational choice.

The biggest obstacle to wider use of RE in North America seems to be reluctance or the apprehension of some clients, engineers, architects and builders. To see more successful RE projects would increase their confidence. The mass proliferation of domestic RE homes in Australia, that aids RE awareness greatly, is unlikely there due to the base low cost and quality of the common conventional lightweight homes often seen there in "tract" housing. Even the Pneumatically Impacted Stabilised Earth (PISE) that David Easton developed in USA could not compete on pure price with the cheap "tract" housing of USA. In Australia the ACCC advises that some 80% plus of external home cladding is fired clay brick. In WA some 95% plus of homes are made of very high quality cavity fired clay brick, with which RE has been able to compete on price, quality, speed of construction and to sometimes exceed on thermal, sustainability, beauty and general "green" credentials. Particularly in Australian areas that are remote, or subject to cyclonic wind or are earthquake areas, the costs can favour rammed earth even more. Similarly on problem clay sites where brickwork must use the more costly articulated masonary construction, yet this comes standard with rammed earth, as no extra cost.

In the writer's view more RE would be built in Australia if more architects, designers, builders etc were totally familiar with it, more engineers had the confidence to readily certify it, more of the unknowns were researched away, thermal approvals were easier, the NCC was improved and big building interests took part in building RE walls so that larger projects could be undertaken with contractual confidence, rather than the small RE contractor (often contractually disadvantaged) being disadvantaged by the big builders (with contractual prowess and only one job to get through in RE) as has so often occurred. If a major outcome of this conference is that more Australian engineers gained the confidence to use (and certify) RE widely then that is a good outcome since the NCC relies on engineers to a significant extent and their confidence in RE will move the RE industry forward a lot.

6 CONCLUDING REMARKS

Thermal advances need to be sped up to provide better modelling, better laws, movement away from the outdated and inappropriate static R value test (a test carried out at fixed temperature) to more dynamic tests and inclusion of full hygrometric modelling which should so advantage RE and so disadvantage many "normal" building materials, whilst giving consumers better comfort conditions with less energy use. Full life cycle analysis should be implemented as it too should better showcase the benefits of rammed earth and also disadvantage many popular "normal" building materials, whilst overall significantly lessening CO_2 emissions.

In places where there is abundant labour and a need for buildings, such as remote communities, homeless areas, disadvantaged and poor people, refugee camps, within volunteer groups, indigenous groups etc, there is the opportunity to use rammed earth on a large scale over wide areas. We, the developed world, can provide the knowhow, equipment and training with labour and materials provided locally. This is a challenge to everyone present. With cement stabilised rammed earth having been used in many huge and successful dams, labelled Rollcrete in USA and elsewhere, there is a great opportunity to use the material in more civil works.

With stabilised rammed earth having been used in many successful building footing, floor, wall and roof situations and unstabilised rammed earth having been used in many successful building floors, walls and roofs, now is the time to use these techniques more widely. Developments in computer cut laminated veneer lumber and plywood and steel and separately in fabric forms will open the door to highly unusual architecture. Rammed earth can peak the "unusualness" since not only is the form/shape totally flexible but also the colours and textures, and over a very wide range. Detailed developments are likely in the field of 3D printing of houses. Mud (liquid earth) homes have already been printed and rammed earth will follow, in time. Using continuous mixing and delivery systems as developed by David Easton, Meror Krayenhoff and the author and continuous forming systems (to Ramtec hunches) and continuous ramming systems currently in use by Martin Rauch there is currently a big opportunity out there for a bold entrepreneur.

The future of RE will significantly be determined by people at this conference who hopefully at conference conclusion will have greater confidence to use more RE. For more than 30 years, many individuals have rammed away often alone, but now we have an emerging synergy. We are all moving onward and upward.

REFERENCES

ASTM (2010). ASTM E2392/E2392M-10e1: Standard guide for design of earthen wall building systems.
Beckett, C., Ciancio, D., Huebner, C. & Cardell-Oliver R. (2014, July 9–11). Sustainable and affordable rammed earth houses in kalgoorlie, western australia: Development of thermal monitoring techniques. In *Proceedings of the Australasian Structural Engineering Conference*, Auckland, NZ.
BG (2014). Architecture. http://biomimicry.net/about/biomimicry/case-examples/architecture/ [accessed: 30/10/2014].
Boltshauser, R. & M. Rauch (2011). *The Rauch House*. BIRK.
Ciancio, D. & Beckett C.T.S. (2013, 19–21 August). Rammed earth: An overview of a sustainable construction material. In *Third International Conference on Sustainable Construction Materials and Technologies*.
Earth Architecture (2014). Earth architecture. http://www.earth-architecture.org/. [accessed: 18/11/2014].
EBAA (2008). *Building with earth bricks and rammed earth in Australia* (Second Edition ed.). EBAA (Aus).
eTool (2014). eTool. http://etoolglobal.com/. [accessed: 18/11/2014].
Hall, M. & Allinson, D. (2009). Analysis of the hygrothermal functional properties of stabilised rammed earth materials. *Building and Environment 44*(9), 1935–1942.
Minke, G. (2006). *Building with earth—Design and technology of a sustainable architecture*. Birkh auser, Basel.
NZS (1998). NZS 4297:1998. Materials and workmanship for earth buildings incorporating amendment no. 1.
Page, A., Moghtaderi, B., Alterman, D. & Hands S. (2011). *A study of the thermal performance of Australian housing*. Priority Research Centre for Energy, The University of Newcastle.
Rael, R. (2009). *Earth Architecture*. Princeton Architectural Press, New York (USA).
SAZS (2001). Sazs 724:2001standard code of practice for rammed earth structures.
Stone, C. & Bagoňa M. (2013). Thermal responses of stabilized rammed earth for colder climatic regions. In M. Kalousek, M. Němeček, and L. Chuchma (Eds.), *Advanced Materials Research*, Volume 649, pp. 171–174.
Walker, P. & Standards Australia (2002). *HB 195: The Australian Earth Building Handbook*. SAI Global Ltd., Sydney, Australia.

Rammed Earth Construction – Ciancio & Beckett (Eds)
© 2015 Taylor & Francis Group, London, ISBN 978-1-138-02770-1

The future and the common ground

David Easton
Rammed Earth Works, Napa, CA, USA

ABSTRACT: Marissa Mayer was the first female engineer at Google. She was responsible for the fun. When she left to become CEO at Yahoo, she put an end to the policy of working from home. In her words, "Some of the best decisions and insights come from hallway and cafeteria discussions, meeting new people, and impromptu team meetings" (Marks 2013). Scientists, academics, engineers, and practitioners, come from as far as it is possible to travel on this planet, finally, in the same room for the ICREC event. ICREC has put an end to working from home. For four days, lab-coated scientists can exchange ideas and experiences with dust-encrusted practitioners, every one committed to a common cause—advancing the credibility of rammed earth as a global solution to durable, healthy, ecological shelter. This paper gives a personal overview of the past and future of rammed earth, with a particular emphasis to the rammed earth activities in North America in the 40 years.

1 INTRODUCTION—THE BEGINNINGS AND THE FUTURE OF RAMMED EARTH

I've been obsessed with rammed earth for forty years, but never has an opportunity such as this presented itself.

I grew up in Southern California in the 1950's. One of my earliest memories was building earth dams to guide the irrigation water in my father's orange grove. The soils of Southern California are primarily sandy alluvium, eroded over millenia from the granite and sandstone of the San Bernardino Mountains. The alluvium was very easy to dig, even for a six year old. I had a shovel put in my hands at a very young age. One of my second early memories was watching bulldozers and steam shovels push over the orange trees and build dams from that same sandy alluvium to encircle the playground that became Disneyland. That's right, my father sold the family farm to a crazy cartoonist from Los Angeles who made his money on a talking mouse. As a show of gratitude for giving up the farm, Walt gave us lifetime free passes to Disneyland. All those impressionable years of my developing youth, spent in the magic kingdom, the happiest place on earth. This very likely molded my adult personality. I spent far too many days in a land where the impossible is made easy and where dreams come true.

When did rammed earth begin? 10,000 years ago in China? 5000 in Mesopotamia? 3000 in Asia Minor? And when did modern rammed earth begin? In 1200 AD in Spain when the Moors constructed the Alham-bra? In 1820 in France with Francois Cointeraux (Lee 2008)? 1840 in the US with Samuel Johnson (Johnson 1806)? 1930 in the US with Tom Hibben and Ralph Patty (Hibben 1941)? In the 1950s in Australia with George F. Middleton (Middleton and Schneider 1992)? Johnson built the Church of the Holy Cross in South Carolina. It's still one of the most significant old rammed earth buildings in the US.

To be sure, the Chinese, the Moors, and other civilizations have kept rammed earth continuously alive—alive but not thriving. Cointeraux, Johnson, Patty, Middleton—they were enthusiasts in their time, but when they died or retired, rammed earth slipped into a sort of architectural dormancy.

I'm going to stick my neck out here and say modern rammed earth began in 1976, when Stephen Dobson built his first rammed earth house in Darwin and I built mine in the California foothills. Quentin Branch was in Arizona, and Patrice Doat and Hugo Houben were in graduate school discovering the widespread effects of Cointeraux's passion for rammed earth.

In 1975, I bought a hundred acres of river-crossed woodland at the end of a paved road in the foothills of eastern California, far enough off the beaten path that I could hide out from building inspectors. Little did I know when I bought that land that the red clay soil under the pine needles would produce beautiful rammed earth walls. One river valley north, and one hundred years earlier, Chinese gold miners found that same red clay soil and built structures that are still standing to this day.

Two little books provided my introduction to rammed earth: Farmer's Bulletin 1500 (Betts and

Miller 1937), published by the US Farm Bureau, and Build Your House of Earth by George F. Middleton (Middleton 1980). We bolted some heavy wooden forms together and tried our hands at pounding earth—sheds, barns, and small houses for me and my neighbors, each one a little better than the one before it, the forms a little less cumbersome, the mix a little more uniform. We discovered the versatility of the pipe clamp and the wide wooden waler, bought a used rammer, an old air compressor and a front loading tractor. We were in business.

In 1981, we came out of the foothills and had the audacity to build the Haywood Winery in Sonoma (shown in Figure 1). Stephen Dobson and Giles Hohnen built the St. Thomas Church in Margaret River (Figure 2)—both surprisingly large undertakings considering how little we knew back then.

But back to the 1970's: What in the world were Stephen, Giles, Quentin, and I thinking? Who in his right mind would believe that you could simply pound dirt into durable shelter—that walls built this way could stand up to wind, weather, and gravity? Who in his right mind would think you could make a business out of such a thing, that people would actually pay for it? What were we thinking?

Figure 1.   Haywood winery in California, USA.

Figure 2.   St. Thomas Church in Western Australia.

That here was an opportunity to support our families and put our kids through college? That we would get rich and successful and launch a global renaissance? Mark Twain once said, "it takes two things to be successful in life—ignorance and confidence." Look at us old timers today. What in the world made us stick with the idea of rammed earth? Was it ignorance, or confidence?

The year was 1979, I was applying for a building permit in a new county in California. I brought my plans and engineering to the office of the chief plan checker, explained that the walls were to be rammed earth, stabilized with cement and incorporated within a concrete post and beam frame, and that he needn't worry because I was a graduate of the Stanford school of engineering. His reaction: "Rammed earth? I never heard of it, but I can tell you right now I don't like it". He wasn't alone. Forty years ago almost no one had heard of rammed earth. And yet here we all are today to celebrate the past successes and to lay out the future of rammed earth.

And what is that future? Many people believe we are on the brink of a global environmental crisis, consuming resources at a lightning pace. During the Carboniferous Period, between the Devonian and the Permian, roughly 350 million years ago, the earth was a vast sweltering swampland. For 60 million years the bodies of billions of dead organisms, buried in mud, and under intense pressure and heat were lithified into hydrocarbons. It has taken 350 million years to build up the hydrocarbon reserves that we industrialized humans have burned up in a mere two hundred. The construction industry is responsible for a large share of this resource consumption. Cement in particular can be held accountable for 5% of all CO_2 emissions. To make cement we incinerate hydrocarbons in order to calcine limestone, which is mostly lithified coral. In the process we release the carbon dioxide that was stored in both the limestone and the hydrocarbons. One ton of CO_2 is released to the atmosphere for every ton of cement produced—not a particularly efficient rate of conversion.

Rammed earth practitioners and researchers should seek out a common ground among them and ask each other this question: *how can we retain the quality of our rammed earth and at the same time reduce our use of cement?* Scientists can guide to improve mix designs and seek out alternative stabilizers. Engineers can develop new protocols that recognize the improvements. Architects can adjust their designs to accommodate the means and the methods established by the practitioners. By doing so, the high visibility projects in Europe, Australia, and North America can be leveraged into expanding rammed earth technology to the places where it can do the most good—building healthy, afford-able, durable housing for the rest of the world.

2 IT IS NOT EASY TO DO WELL

When I was in my mid twenties, two people pro-vided me with inspiration to walk the rammed earth path. One was David Miller, a country law-yer from Colorado who when he wasn't arguing water rights cases between the wheat farmers and the sheep ranchers, was a globe trotting rammed earth historian and an owner builder. The other person was Wayne Dunlap, a geologist from Texas A & M who wrote a milestone manual, Hand-book for Building Homes of Earth (Wolfskill et al. 1963). I met them both at the first national con-ference on Earth Building in Albuquerque New Mexico in 1980. David was one of the very few who kept the rammed earth dream alive between the depression era years and the mid 1970's when it was "re-discovered".

Wayne Dunlap was one of the most inspirational college professors I ever listened too. I wish I could have taken a full course from him rather than the one short lecture. In his will, David bequeathed me his library of photos and rammed earth research. You could put the little handbook Wayne wrote in your coat pocket, travel anywhere in the world, and feel confident selecting the best soil to build rammed earth. The handbook made it seem so simple.

Simple was the operative word at the start. Let me tell you those were the days. Simple forms, sim-ple buildings. Grab a bucketful of soil from the footing excavation and beat it so hard it stands up on its own. I remember the first time I did this—set a form, filled it with pounded earth, and stripped the form—I could hardly believe my eyes. Could this be real? Why doesn't everyone build this way? This is so simple! I was mesmerized, transfixed, I was as smitten as Francois Cointreaux. My life's work was laid out in front of me as clearly as Dor-othy's yellow brick road leading to Oz.

But Wayne's handbook contained a warning: Number one on his list of the disadvantages of building with rammed earth: "it is not easy to do well". He was right. It is not easy to do well. The yellow brick road to Oz wasn't as well-paved or as clearly visible as I first thought.

The Myth of Sisyphus is a philosophical essay written by Albert Camus in 1942 (Camus 2000). The original Greek myth tells us that Sisyphus was condemned to repeat forever the same meaningless task of pushing a boulder up a mountain, only to see it roll down again. Camus maintains, however, that "the struggle itself is enough to fill a man's heart. One must imagine Sisyphus happy". Have

I been happy pushing a rammed earth block up a hill for forty years? Absolutely. Remember, I spent my youth in a land where the impossible is made easy and where dreams come true.

There is a road in California, the original road, as a matter of fact, but back then it was more like a dirt path. It runs from the southern border with Mexico for six hundred miles, all the way to Sonoma, an hour north of San Francisco. The road is called El Camino Real—the King's High-way. In the 1940's it was paved with macadam, a type of rammed earth in which layers of small bro-ken stones are compressed.

Sometimes macadam is mixed with a bitumen, but not always. That macadam has mostly been replaced with concrete by now, but along this road are scattered twenty-one two hundred and fifty year old adobe missions. My father took us to every one of those missions (one of those presented in Figure 3). Fourth graders in California public schools make models of them. The missions are part of California's heritage. You might say adobe was the original California building material.

The adobe missions have provided a valuable field laboratory over the years. Every one of the bell towers and tall gable end walls has tumbled down in one earthquake or another—1812, 1827, 1857, 1923—but the low walls survived. Forensic engineers were able to look at the slenderness ratios and extrapolate safety standards. The important information garnered is that slenderness ratios under 4.5 can survive earthquakes, even with unit compressive strengths of 0.25 MPa (30 psi).

I would have thought building rammed earth with slenderness ratios below 4.5 and compressive strengths well above 0.25 MPa would have satisfied the building department. "Just look at all of those field tests verifying the numbers, each of them over two hundred years old". No luck. Sometimes it feels like trying to convince a building official is like being stuck in geologic time.

Figure 3. Mission of San Francisco Solano, California, USA.

Another quote from Mark Twain. "Whenever you find yourself on the side of the majority it is time to pause and reflect". You had to be on the opposite side of the majority to be a rammed earth builder in the 1970's, certainly in California where earthquakes put the shakes in the heart of every structural engineer and code official. I could not have picked a worse place to revive rammed earth than California if I had tried. Think of Sisyphus pushing the stone up a cliff—in geologic time. The County of Los Angeles is so obstinate we still haven't been able to crack the code there in forty years.

3 WHAT IS RAMMED EARTH?

Perhaps even before we start building our macadam highway, we should agree on one important common definition. What is rammed earth? Is it a method of building structural walls on site in which soil, aggregates, sometimes with a binder are beaten down in layers until each is hard? Or is it no longer considered rammed earth if the mix design is not pure earth, but crushed aggregate mined in a quarry?

Is rammed earth only pure compacted soil or is it any aggregate pounded into a monolithic wall, whether or not blended with stabilizer? Is it the act of ramming, or the composition of the earth? Is a poured earth wall rammed earth? Is a shot earth wall rammed earth? Is a wall of compressed earth blocks rammed earth? What defines rammed earth? The material, the method, or it's monolithic character? Does cement stabilization change the character of the wall so much that we can no longer call it rammed earth?

To many people, rammed earth must be made of pure earth. As soon as cement is added, the product becomes something different—it loses its hygroscopic properties, its clay properties are gone forever, and most importantly it is only marginally sustainable: remember using 10% cement stabilization in a 400 mm wall releases more CO_2 than a cast in place concrete wall.

I believe there are some builders and researchers who would argue that those adding cement are not building true rammed earth at all, but an ugly sister relegated to a sort of marriage of convenience. A marriage of convenience, is *a marriage contracted for social, political, or economic advantage rather than for mutual affection*; broadly: a union or cooperation formed solely for pragmatic reasons. Was the union of earth and cement contracted solely for pragmatic reasons: to ease the pathway, to build a wide paved road, an El Camino Real, leading to building department approvals and public acceptance? Or was the marriage of earth and cement contracted with an expectation of mutual affection in order to improve the strength and performance of rammed earth in general and to allow for the use of otherwise marginal soils?

4 CEMENT-STABILIZED RAMMED EARTH

Cement was being used as a stabilizer for rammed earth as early as the 1940's. Earlier builders had tried urea, manure, fiber, lime, bitumen and other admixtures. Portland cement was the best of them all, and this was long before the environmental effects of calcining limestone and releasing CO_2 were an issue. Cement, after all, was a very effective and affordable glue, capable of improving the strength and durability of raw rammed earth by a factor of five.

The question now might be: since we have worked together for forty years to create acceptance for rammed earth as a modern building medium, can we begin to retrace our steps? To go backwards, as illogical as that sounds. Can we remove the cement from our mix designs or at least cut back, from 10% to 5% or 2.5%? Can we find an alternative to Portland cement that will give us strength, durability, resistance to erosion and still maintain the credibility we have so patiently acquired? Can we find common ground?

The researchers at Watershed Materials, working with the support of the National Science Foundation ($740,000 Phase II Small Business Innovation Research grant from the National Science Foundation), are in the second phase of a testing program and now obtaining strengths up to 41 MPa (6000 psi) using geopolymers to replace Portland cement.

In my opinion, the battle around the reduction of elimination of the use of cement is as much public policy as it is structural safety. In the US, we are compelled to achieve compressive strengths higher than actual design calculations would require. There are layers of safety imposed on structural design that force us in this direction, some by structural engineers others by policy makers, some by the fear of lawsuit. Variations in how to interpret the code, safety factors, and design guidelines, especially in California, can lead one engineer to require a compressive strength of 5.5 MPa (800 psi) where another engineer will only feel confident with two times those strengths; 11 MPa (1600 psi) and higher. Some builders and engineers try to achieve strengths of 17 MPa (2500 psi), equal to that of cast in place concrete, as a pathway around the code.

Why do the world's codes differ so radically on the perception of what is safe rammed earth—

0.25 MPa (30 psi) in some countries, 17 MPa in others? In soils, it takes a minimum of 10% Portland cement to achieve strengths of 17 MPa, less cement in crushed aggregate. What this means, distressingly, is that there is nearly twice the cement in a 400 mm stabilized earth wall than there is in a typical concrete wall, and every pound of cement calcined generates nearly a pound of CO_2.

This is the dirty little secret we are not sharing about rammed earth. It's akin to the myth of Pandora. Today the phrase to "open Pandora's box (it was actually a jar)" means "to perform an action that may seem small or innocent, but that turns out to have severely detrimental and far-reaching consequences". Pandora removed the lid and all manner of evil escaped and spread over the earth.

In the instance of stabilized rammed earth, it isn't that evil will spread over the earth, but what will happen is that our claims that rammed earth is an environmentally benign, even beneficial wall assembly will be seriously challenged. Yes it saves on lumber, drywall, and paint; it outlasts other wall assemblies and requires far less maintenance, but cement, ordinary Portland cement is, after all, responsible for 5% of all the world's CO_2 emissions.

Let me share with you how I got addicted to cement. I confess I am one of the worst offenders. I confess I fell under the spell that stronger is always better. Cement made me feel invincible. I look back on my past and I can see when this addiction started. I began building in the 1970's, taking earth from the site and pounding it into simple, not very elegant walls—fast, inexpensive, but somewhat crude. Very little energy input other than human carbohydrates.

As opportunities grew for our struggling company, the marketplace—architects and clients—wanted the magic and the muscle of rammed earth, but wished it were not so crude. Building officials and engineers wanted it not so unrefined and difficult to specify. Here came our first big price jump—quarry materials, steel reinforcing, better forming methods, slower more careful work, higher wages, diesel fuel, and the demon cement. Only a little at first, but gradually we became heavy users, lured by the vision of code approval.

4.1 Pneumatically Impacted Stabilized Earth

In the 1980's we invented PISE—high pressure air delivery. PISE for Pneumatically Impacted Stabilized Earth (Figure 4). No shovels, no ramming, half the formwork. We were working on a very large construction site for a wild animal theme park. We were building rammed earth termite mounds, and next to us was a gunite crew shooting concrete against a dirt bank, then carving it and

(a) work in progress

(b) final construction

Figure 4. Examples of Pise—Pneumatically impacted stabilized earth.

painting it to look like earth. I studied their equipment for a few hours and got to thinking maybe we could shoot rammed earth that way.

The early attempts were troubling. I'd say 90% discouraging, but there was 10% hope, the same hope left in the bottom of Pandora's box. To continue working to perfect PISE called once again on that unique combination Mark Twain identified: ignorance and confidence. We persevered to bring PISE to the marketplace.

PISE was fast. We shot thousands of cubic yards of PISE throughout the wine country of northern California in the 80's and 90's. Wine makers especially liked it because it was an inexpensive way to get great thermal storage. But pise needs even more cement than rammed earth due to the higher water contents. Over the years, as cement and diesel prices went up and concerns over global warming grew, my passion for PISE waned. I could no longer justify burning 100 gallons of diesel fuel to power the big air compressor to get 1000 square feet of finished wall. It didn't pencil out, financially or ecologically. I had to kick the habit.

The good news, I'm proud to say, is that I am on the road to recovery. Our testing programs are verifying that we can attain the necessary strengths at reduced cement ratios, and for our most recent two rammed earth installations we dropped our cement use by 30%.

5 RAMMED EARTH AND BIOPHILIA

This raises a question: What is it that makes rammed earth so attractive, so alluring, so captivating? What exactly is the magic? Is it simply the hygroscopic ability of raw earth to maintain optimum humidity levels within a space? Or is it the way thick earth walls can soften sounds and provoke a sense of calm? Perhaps they capture the essence of biophilic design, that the earth walls provide a source and sense of connection to the natural world, distilling natural materials to their elegant simplicity and rightness of fit.

The recent interest in biophilia—architecture to connect people with nature—could not find a better mascot, a better poster child than rammed earth. A thick, strong earth wall acts like a filter, excluding the noise and the stress that is outside, creating a positive, beneficial environment within. It's pure and simple.

I have my own idea of what makes rammed earth so endearing and mesmerizing. It links us to geologic time. The first law of thermodynamics states that energy can be transformed from one form to another, but cannot be created or destroyed.

I believe this construct can be applied to raw earth. 4.6 billion years ago, when the magma that would become planet earth was cooling for the first time, rock began to form. We call this first generation rock precambrian—a period in the earth's history that began 4.6 billion years ago and extended to the beginning of the Cambrian era 540 million years ago. Geologists have been forced to lump 80% of all the world's geologic metamorphosis into this one vast period of time because there was no life, no fossil record prior to 546 million years ago, and hence no accurate way to date the origin of the rock. In a sense rock goes through a life cycle. It grows, either from heat or pressure, and it dies, from weather, grinding, or leaching, only to form again.

The individual grains of gravel, sand, silt and clay used to build rammed earth walls are as old as the earth itself, the product of the big bang, the molecules that from nothingness fused to become the planet we call home, and which we rammed earth builders transform into shelter, into home.

Yes, we rammed earth builders have mastered a technology, and we've done so in such a way as to capture the nature of rock, the natural, made-by hand character, the wabi sabi, of the material itself, it's incredible age and it's tenacity—its endearing quality. After experiencing a building with thick earth walls, there is no going back.

This is what keeps us pushing that mammoth stone of rammed earth up the hill, but is it the ancient precambrian rock composition or is it purely biophilic determinism? Is it tenacity or obstinate contrarianism?

6 CONCLUSIONS

I think we're all big wave riders, on the crest of a green building revolution, and it's either confidence or ignorance that keeps us here. Will rammed earth prove to be the most ecologically responsible of all wall systems—bringing safe, healthy, affordable housing to people in need, while at the same time giving, modern architects a massive new materials pallet?

Wayne Dunlap warned me, and we all know, it is not easy to do well. But let's all stand committed, from this point on, to doing it right.

REFERENCES

Betts, M.C. & T.A.H. Miller (1937). *Rammed earth walls for buildings. Farmers Bulletin No. 1500.* U.S. Department of Agriculture, Washington D.C. (U.S. Govt. Printing Office).

Camus, A. (2000). *The myth of Sisyphus.* Penguin Group, London (UK).

Hibben, T. (1941, October). Rammed earth construction. Technical Support Circular No. 16, Supplemental No. 1. Technical report, National Youth Administration, Washington, DC.

Johnson, S.W. (1806). *Rural economy.* I. Riley and Co., New Brunswick (USA).

Lee, P.Y. (2008). Pisé and the peasantry: François cointeraux ´ and the rhetoric of rural housing in revolutionary paris. *Journal of the Society of Architectural Historians 67,* 58–77.

Marks, G. (2013). *Why Marissa's right.* Forbes.

Middleton, G.F. (1980). *Build your house of earth—a manual of earth wall construction.* Compendium/Second Back Row Press.

Middleton, G.F. & Schneider L.M. (1992). *Bulletin 5: Earth-wall construction* (4th ed.). CSIRO Division of Building, Construction and Engineering, North Ryde, Australia.

Wolfskill, L.A., Dunlap, W.A. & Gallaway B.M. (1963). *Handobook for building homes of earth.* Rammed Earth Institute, International.

Rammed Earth Construction – Ciancio & Beckett (Eds)
© 2015 Taylor & Francis Group, London, ISBN 978-1-138-02770-1

Earthen construction: A geotechnical engineering perspective

C.E. Augarde

School of Engineering and Computing Sciences, Durham University, UK

ABSTRACT: While earthen construction has been an activity undertaken by Man for thousands of years it is only less than a decade that the material has been considered in the framework of soil mechanics. Much of modern earthen construction is based on the assumption that one is dealing with a homogeneous, isotropic material which is elastic, to all intents and purposes, until it suddenly fractures, an approach which bears the hallmark of thinking along structural engineering lines. This is quite reasonable as we are constructing using manufactured materials, something that we usually do in structural engineering, but quite different to the approach taken in geotechnical engineering, where we are usually dealing with a natural material which we may disturb a little. Earthen construction seems to fall somewhere between these two sub-disciplines of civil engineering; the material is manufactured but it is still a particulate soil-based material. This paper makes a link between concepts now routinely used in geotechnical engineering and earthen construction materials. The paper is illustrated with many examples of research undertaken over the past decade at Durham with this approach.

1 INTRODUCTION

Despite the potential environmental benefits, it seems difficult to imagine a large scale uptake of earthen construction for routine construction of parts of buildings in the UK given the dominance of the industries which produce the conventional alternatives, of fired brick and concrete. If any major progress is to be made on this front I contend that we have to "play ball" with the construction industry and present earthen construction in similar terms to conventional materials. An approach found commonly in the UK is to reject attempts to modernise or apply scientific rigour to understanding earthen construction materials, which is fine if you are happy for the environmental benefits of earthen materials to be visited only upon those well-off enough to afford bespoke construction. Without scientific understanding of how these materials behave, feeding into design procedures, it is unlikely they will become easily specifiable alternatives on a scale where real environmental benefits accrue to all. This is not a criticism of the many good pieces of engineering research that have been carried out on earthen materials. The aim of this paper is to point out an alternative that might be considered.

Geotechnical engineering is usually solely concerned with building things in, or on the ground (i.e. in or on the subsoil, the earth in earthen construction). In some situations geotechnical constructions are built *with* the ground, for instance an earth dam is built using subsoil excavated, transported and compacted into place, but this use of earth differs from its use in earthen construction.

My background is in conventional civil engineering, having become a UK Chartered Engineer between a first degree and returning to academia in the mid-1990s. I first became aware of earthen construction in the early 2000s through Prof. Chris Gerrard, of the Dept of Archaeology at Durham. He had a long-term interest in medieval buildings in Spain (principally Aragon) constructed from rammed earth and adobe, and had some concerns over the structural stability of one building in particular. Over a number of years working with Chris, and sharing a PhD student, I have come to see the variety and number of heritage earthen structures there are, even in Europe and the UK.

What struck me about this experience as an engineer was that here was a natural building material, clearly durable in the right conditions, which could have the green credentials for a new material in construction. What I also realised was that, as far as I could see, the materials had not been viewed from a geotechnical perspective. Since the mid-2000s I have tried to rectify this through a number of small scale projects, a book and journal publications. It is clear that the topic of earthen construction is becoming of greater interest to geotechnical engineers, and this is a good time to survey this perspective, here in this paper.

2 SOME SOIL MECHANICS

2.1 *Effective stress*

With apologies to those of you who are either geotechnical engineers already, or who remember your university soil mechanics, I am going to run through some basic concepts routinely used by geotechnical engineers and will link them to earthen construction materials.

In this paper, and in much of my research to date, I have focussed on strength as opposed to stiffness. The latter is usually more important, if we consider the difference between an ultimate limit state of collapse and the service ability limit state of deflection, and is also much easier to explore with soils. Soils generally have almost no linear elastic response (although this is usually built in to all constitutive models for soils, i.e. the mathematical model linking stresses and strains) and predicting deflections of geotechnical constructions is notoriously difficult to get right.

Recognising that soils are particulate leads to the assumption of frictional behaviour, i.e. friction between individual grains leads to macro-level strength properties, e.g. the ability of soils to sit in conical heaps, and failure is due to limiting shear strains being reached (rather than grains crushing, although they do that too). Considering a plane in a soil, the shearing stress τ on the plane is a function of the normal stress σ across the plane, linked by a coefficient of friction. In soils we usually write this as.

$$\tau = \sigma \tan \phi \tag{1}$$

where ϕ is referred to as the angle of friction for the soil. This simple model does not however lead to predictions of failure unless the normal stress changes, and in practice we see the failure, of geotechnical constructions when water conditions change. Terzaghi's principle of effective stress incorporates the role of water into the basic friction model. Terzaghi stated that all the behaviour of soils is due to changes in the *effective stress* σ' which he defined as

$$\sigma' = \sigma - u \tag{2}$$

where σ is referred to as the *total stress* and u is the *pore water pressure*. By this simple approach we can have failure due to a decrease in effective stress, either via a decrease in total stress or an increase in pore water pressure. Physically, it is easy to see that if the water in the soil is pushing the particles apart, then they will have reduced shear strength.

Importantly, Terzaghi's effective stress relates only to saturated soils, i.e. where all the voids between particles are filled with water. If the soil has voids filled with air and water then this is referred to as an unsaturated or partially saturated soil, and this seemingly innocuous change has a major effect on mechanical properties.

2.2 *Unsaturated soil mechanics*

In the majority of cases, geotechnical engineers assume full saturation of the soils they work with and make assumptions about strength and stiffness on that basis. While this has clearly "worked" insofar that things get designed and built and tend not to collapse, there is an increasing recognition that many of the soils engineers work with are unsaturated, maybe because they lie in the vadose zone above the water table, or they are subjected to wetting and drying cycles due to weather and climate, and that there are advantages to taking this into account.

The principal benefit of partial saturation is the presence of *suction*, which can be regarded as a negative pore water pressure which pulls particles together. If you insert a negative u into Eqn 2 you can see that increasing suction means increasing effective stress (using Terzaghi's measure), and hence increasing shear strength. There is some disagreement in this area, as some cannot imagine water taking a tensile stress, and at high suctions it is likely that some other mechanisms (maybe to do with adsorbed water on particle surfaces is at play). It is however undeniable that in experimental testing we observe that the shear strength of soil increases with suction.

We measure the degree of saturation S_r as a dimensionless value between 0 and 1 as

$$S_r = V_w / V_u \tag{3}$$

where V_w is the volume of water in a sample and V_u the volume of voids, and an important hydraulic property of unsaturated soils is the Water Retention Curve (WRC) which links degree of saturation to suction. A typical WRC is shown in Fig. 1 and it is important to note that experimentally the relationship is hysteretic, differing if the sample is wetting or drying. The shape of the WRC is closely linked to the sizes of pores in the sample, the Void Size Distribution (VSD). Large pores can onoy sustain low suctions, and vice versa. So a clay material, with very small pores can carry a very high suction and this is often what is mistaken for an apparent cohesion.

Many natural phenomena can be explained with this idea of the link between suction and partial saturation, e.g. rainfall induced landslides occur because high suctions are holding the slope together, until rain comes, the slope wets and the suction decreases or disappears.

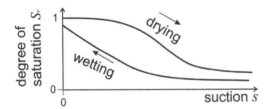

Figure 1. The water retention curve (semilog axes) in unsaturated soil mechanics (from Augarde 2012).

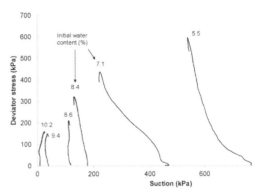

Figure 2. Results from unconfined compression tests on unstabilised RE including suction measurements. (from Jaquin et al. 2009).

Unsaturated soil mechanics is still very much a realm for research rather than application in practice. Key areas are the development of constitutive models many of which use variants of Terzaghi's effective stress and suction to determine both strength and stiffness, and the development of laboratory and field testing equipment to measure suction.

2.3 *Applying geotechnical thinking to earthen construction materials*

Given the background in soil mechanics above, where do earthen construction materials fit in? Clearly the basic materials are the same, excepting where stabilisers are added, although geotechnical engineers sometimes stabilise natural soils using cement or lime to improve settlement characteristics. The major difference is that earthen construction materials are manufactured and tend to be dried out to a much greater degree than any natural soils. They therefore operate at much higher levels of suction than found in natural soils which means they are stronger, but also the low water contents lead generally to brittle behaviour.

The first publication that we are aware of that makes the link between suction and the strength of earthen construction materials is the conference paper of Gelard et al. (2007), which came at the same time as the group at Durham was carrying out the experimental tests which later appeared in Jaquin et al. (2009). These comprised constant water content, unconfined compression tests on unstabilised Rammed Earth (RE) samples where the suctions were measured during testing using devices called tensiometers. A clear link was shown between suction and strength, an example of which is shown in Fig. 2 where shear strength (measured here as deviator stress) is plotted against suction as the tests proceed. The suctions measured here are much lower than those found in the field after drying, however the principle is clear. In fact the idea that a component of strength in earthen construction comes from suction is quite obvious, as the means by which beach sandcastles gain enough strength to stand up.

It appears then that there is mileage in considering earthen construction materials as *manufactured unsaturated soils,* and to develop constitutive models of mechanical and hydraulic behaviour from a geotechnical point of view. There is evidence that this approach is gaining interest, via recent papers (e.g. Nowamooz & Chazallon (2011)) and a keynote at the most recent International Conference on Unsaturated Soils (Gallipoli et al. 2014). In the following sections I will set out some examples of research work undertaken at Durham over the past 9 years involving earthen construction materials considered from this geotechnical point of view. Due to space limitations I will not be covering our work on heritage structures and materials, mainly the work of Paul Jaquin, Chris Gerrard and myself, (e.g. Jaquin (2008); Jaquin et al. (2008), Jaquin & Augarde (2012)).

3 SOME GEOTECHNICAL INVESTIGATIONS OF EARTHEN CONSTRUCTION MATERIALS

3.1 *Fracture*

Insitu earthen construction materials characteristically fail in a brittle manner and it is therefore natural for this to be a subject of investigation. Fracture mechanics here (as for other materials) is concerned with fracture initiation and propagation and, of the little scientific work published on this, most have adopted linear elastic fracture mechanics principles as a start (e.g. the work presented in Brune et al. (2013) on Roman mortars). Some ideas on appropriate tests for fracture testing for earthen construction materials have been taken up in Corbin & Augarde (2014) which describes the development and use of a wedge splitting device for use with earthen construction materials (Fig. 3), and an example of a fractured stabilised RE specimen using this rig is shown in Fig. 4. With this device one can obtain reliable and repeatable fracture energies for these materials.

3.2 *Fibres and other reinforcements*

Fracture inhibition in brittle materials usually means reinforcement in tension, and fibre-reinforcement is of course a key feature of many earthen construction materials, e.g. the straw in adobe bricks and cob is a form of tensile reinforcement. It is an intriguing question to consider what role these fibres play in terms of water storage and distribution, and the nature of the bond between the fibres and the surrounding material.

Investigations have been carried out at Durham on the properties of fibre-reinforced mixes at the macro scale and also the fibre/earth bond itself. Corbin & Augarde (2014) demonstrated the major change in fracture behaviour between unreinforced and reinforced stabilised RE and also the increase in Unconfined Compressive Strength (UCS) with wool reinforcement (an example plot

is given in Fig. 5). Investigations of the fibre-earth bond behaviour are presented in Readle (2013) who developed a new test rig to carry out pull out tests on samples of earthen materials. In this study Readle determined pull-out loads using a jute fibre embedded in an unstabilised RE mix. Water content, fibre embedment length and dry density were all varied. One example results plot is presented in Fig. 6, showing the effect of water content for a 50 mm embedment, i.e. at high water contents, typically close to initial placement, the bond strength is much lower than after drying, but then the difference between 3% and 7% water content is negligible, so small varaitions due to wetting and drying in the filed are probably not important. Also clear from this plot, and in many of Readle's results, is the presence of peak and residual strengths, potentially unsafe for design, and a feature of overconsolidated soils in saturated soil mechanics. This behaviour is thought to be associated with dilation of the earth material increasing bond strength initially, followed by frictional failure. The scope of the study was extended recently to stabilised materials in Coghlan (2014). Clearly, single fibre studies have to be scaled up to the macro-material case but these

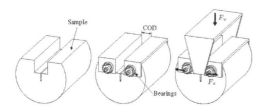

Figure 3. Wedge splitting device for obtaining the fracture energy of earthen construction materials (from Corbin & Augarde (2014)).

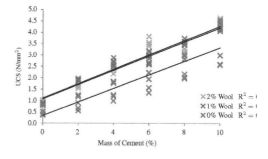

Figure 5. The effect of wool reinforcement on the UCS of stabilised rammed earth samples (from Corbin & Augarde (2014)).

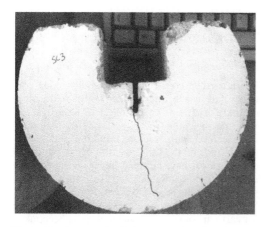

Figure 4. An example of a fractured stabilised RE specimen using the fracture rig.

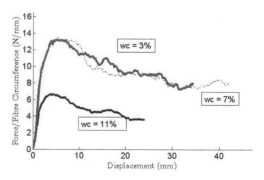

Figure 6. Example of results for pull-out tests on single fibres in unstabilised RE (from Readle (2013)).

findings are interesting for revealing mechanisms of failure. The behaviour of fibre-reinforced soil has been studied by the geotechnical engineering community (e.g. Zornberg (2002)) however these studies are usually for natural soils at much lower compaction levels and higher saturations than the conditions found in earthen construction.

Retaining structures and bridge abutments are routinely designed using "reinforced soil" where strips or grids made from thin plastics materials are incorporated in the soil during construction to provide tensile reinforcement. Could this be used for earthen structures? There are examples of the use of "geogrids" to protect earthen structures from seismic damage but this is often via an overlain sheet retrofitted, rather than appearing in the original build. Howard (2007) presents results from a suite of tests on unstabilised RE in a large shear box (a fairly standard piece of geotechnical engineering test equipment). Some photos of the procedure are shown here in Fig. 7. Howard created samples incorporating geogrid placed during compaction of the RE mix, perpendicular to the layers. He showed that a major increase in interlayer strength, and more importantly ductility, could be obtained with this method and speculated that it could be used to provide protection to seismic loading in new build earthen structures. A large geotechnical shear box was also used to investigate the shear strength of cob in Bargh (2010). Of the many interesting findings from this study (possibly the first time cob had been tested this way) was the ductility change with the straw in the cob as reinforcement. Figure 8

(from this study) shows that plain soil mix tends to exhibit peak and residual strengths (as seen earlier in this paper) wheras cob (with the addition of straw reinforcement) has a safer non-peak response. This study also used tensiometers to measure suctions on the shear plane.

3.3 Testing of earthen construction materials

It has long been known that getting the water content right at the point of compaction is vital for the production of resilient unstabilised rammed earth (Houben & Guillard, 1994; Walker et al. 2005). In the UK and elsewhere, extensive use has been made of empirical tests, principally the "drop test", a simple on-site test procedure whereby the right water content is judged by dropping balls of wetted RE mix and observing how the balls fracture when they hit the ground. In some cases the test procedure states the height of the operative, in others it does not. In Smith & Augarde (2013a) a study was conducted using a large number of engineering undergraduates to carry out drop tests, the aim being to see if by following the drop test instructions, trained and untrained persons could arrive at repeatable water contents for a "good" mix. The results were clear, that the test lacked repeatability and accuracy, in comparison to standard compaction tests. Fig. 9 shows a plot of % error in predicted water content against operator height, indicating the wide scatter of results.

The tensile strength of earthen construction materials is, naturally, often hard to obtain,

| Construction | Drying | Shear plane |

| Shear plane top removed | Broken geogrid sheet | Broken geogrid sheet |

Figure 7.   Large shear box testing of rammed earth samples with geogrid.

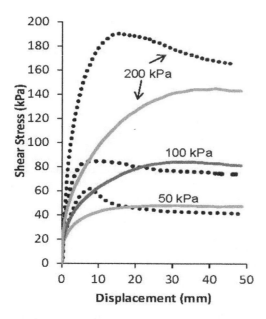

Figure 8. Shear box test results on plain soil (broken lines) and cob (full lines) (Bargh 2010).

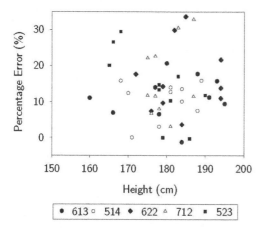

Figure 9. Errors in water contents judged by the drop test for RE. Mix designation from Hall & Djerbib (2004), plot from Smith & Augarde (2013a).

especially for unstabilised materials. The Brazilian test is widely used in rock mechanics and comprises the compressive loading of a disc of rock across a diameter. The disc fails in tension via a crack linking the load application points and, assuming elastic behaviour, one can obtain the average tensile stress (and hence tensile strength) along the fracture. Beckett (2011) includes work showing the applicability of this simple test to earthen

construction materials, concluding that samples of 50 or 100 mm diameter yield the most convincing results. In later work (Beckett et al. 2014) use is made of this test in a study of the effect of salinity in the pore water on the tensile strength of RE. Figure 10 shows one set of results for a mix with silty clay, sand and gravel in proportions 2-7-1 respectively. For this mix the effect of salinity in the pore water is at odds with previously published work and an explanation (contained in detail in the paper) is based on the low clay content of this particular mix. This study has interest outside of earthen construction materials, in agricultural soils in areas at risk of seawater intrusion. Results from Brazilian testing applied to RE (in a slightly different format) appear also in a recent paper by Bui et al. (2014).

3.4 *Investigating microstructure*

One of our key current investigations concerns the microstructure of earthen construction materials, by this we mean the balance between the particle size distribution and the range of pore sizes (the VSD). Two mixes, having the same particle size distribution, can lead to materials with two very different strengths, as part of the recipe is the compaction applied. We have already seen above the importance of the VSD on the water retention and hence suction characteristics for unsaturated materials, and it is compaction that creates the VSD. Various other researchers have attempted to link PSD to material properties, as the PSD is easy to obtain, however from a scientific point of view it must be better to gain an understanding of the VSDs and then link that to the intrinsic material properties observed macroscopically. It is my contention that this is the way to devise design methods for these materials in the future.

Methods for determining the VSD of a soil sample have a long history, with Mercury Intrusion

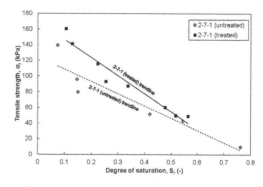

Figure 10. Tensile strength of RE with and without saline pore water (from Beckett et al. (2014)).

Porosimetry (MIP) being an established technique (and not as dangerous as it sounds!) However MIP works with very small samples of soil and is not suitable for earthen construction materials where there is a wide range of particle size.

For this reason, attention has switched, in the geotechnical community at least, to the use of non-destructive techniques which can deliver information on the internal structure of porous media, and the main player here is X-Ray Computed Tomography (XRCT). The technology for XRCT is now very simple and is shown at its most basic in Figure 11. A very low power x-ray source fires rays in a cone through a sample and the attenuated radiation is captured by a detector (really a camera). The sample is then rotated and the x-rays fired again, and so on. 2D images are collected which can be post-processed into 3D "volumes" where the different material densities show up as different brightnesses. The great advantage of this method is that samples do not require any special preparation in advance of scanning, and while the machines themselves are expensive, scans can be obtained relatively easily and quickly.

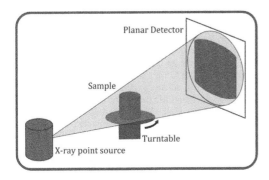

Figure 11.   X-Ray Computed Tomography.

A Zeiss Versa 410 machine has been operating in Engineering and Computing Sciences at Durham since 2013, with earthen construction materials being one of the key materials scanned. Initial work using a machine at Nottingham University, UK, (Beckett et al. (2013)), and a desktop machine (Smith et al. (2014)) has been built on more recently as described in Smith & Augarde (2014), where the potential for the use of XRCT for scanning soil mixtures is examined. There is a conflict in XRCT scanning between wishing to obtain the highest resolution and the largest area of coverage. One can rarely achieve both, and with a compacted material with a range of particle sizes (like a RE mix for instance) one cannot see right "down to the clay". Instead a pragmatic approach must be adopted where sample size is chosen to balance the capabilities of the XRCT machine and the desire for representative samples (i.e. a very small sample will scan well but is unrepresentative of a mix where there could be large sand and gravel particles present). Three sections from scans undertaken for different sample sizes (cylindrical 12, 38 and 100 mm dia.) are shown in Figure 12 where one can see the detail revealed by the scanning. The conclusion of this study is that the optimum choice is the 38 mm dia. sample, balancing the XRCT issues stated above with the need for ease and reality of lab testing.

Figure 13 shows the type of information one can only obtain from scanning. Plots of VSDs in 38 mm diameter samples of a RE mix (denoted 30*:60:10[0.9] using the classification of Smith & Augarde (2013a)) before and after unconfined compressive loading are shown (D = detailed scan; T = top of a compaction layer; B = bottom of a compaction layer; F = full sample scan). The volume of voids in the sample increases during loading, due mainly to cracking in the sample. Work is currently ongoing attempting to link the VSDs

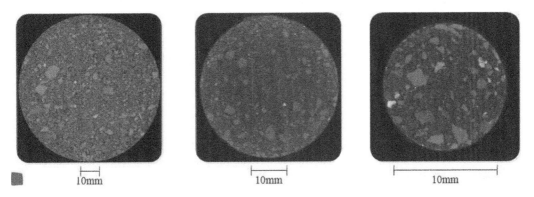

Figure 12.   XRCT scans of cylindrical rammed earth samples of different diameters (100, 38 and 12 mm) showing microstructure at different scales (from Smith & Augarde 2014).

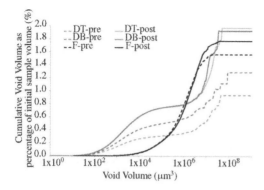

Figure 13. VSDs for samples scanned before and after compression loading.

observed with the water retention curve and hence to suction as one of the sources of strength.

4 CONCLUSIONS

A geotechnical perspective on earthen construction materials has been presented here, illustrated by some of the work in which I have been involved over the past decade. Looking ahead, from the geotechnical research point of view it is clear that researchers in the unsaturated soils community will find increasing interest in these materials. The key challenge is developing constitutive models, as these materials are elasto-plastic and usually brittle. Earthen construction materials are a long way from the near-saturated/low suction wetter soils for which constitutive models have been developed to date. The search for models that can tackle the behaviour of natural brittle soils (often desiccated) is a major line of work as it is. For stabilised earthen construction materials it seems likely that the overlap with weak concrete material modelling is an area ripe for investigation. There is a gap between the soil mechanics models and those for bonded manufactured materials such as concrete, despite there being natural examples of bonded or cemented soils. Earthen construction materials fill the gap.

As will be evident above, students have carried out much of the day to day research work, and it continues to surprise me how many students are keen on this type of project. Some are motivated by the lack of computational work (which is my other main research area as an academic) but most are fascinated by these materials and welcome the opportunity to do lab work in an area for which there is relatively little "out there". The projects have also been good candidates for outreach activ-

ities, and in 2009 a press release on our findings regarding suction was taken up by many newspapers and websites. It is to be hoped that this interest, from students and the public is maintained in the future as this will help the aim set out at the start of this paper, to modernise some attitudes to earthen construction.

ACKNOWLEDGEMENTS

The majority of the hard work in the lab which has led to the research outcomes highlighted above have been down to a small army of students and I would like to record my thanks to: Paul Jaquin, Chris Beckett, Jonathan Smith & Andrew Corbin (PhD students); Tom Horncastle, Tom Howard, Doug Readle, Sarah Coghlan, Lucy Wolf, Hazel Short, Steven Perkins, Josh Bargh & Matthew Woodhams (MEng students). Technical staff have also been a vital part of the team (Steven Richardson and Tom Beale).

REFERENCES

Augarde, C.E. 2012. Strength and mechanical behavior. Ch 8. in Hall et al. (eds) *Modern Earth Buildings*. Cambridge: Woodhead Publishing.

Bargh, J. 2010. *Geotechnical properties of cob*. MEng Dissertation, Durham University, UK.

Beckett, C.T.S. 2011. *The role of material structure in compacted earthen building materials: implication for design and construction*, PhD thesis, Durham University.

Beckett, C.T.S., Hall, M.R. & Augarde, C.E. 2013. Macrostructural changes in compacted earthen construction materials under loading. *Acta Geotechnica*, 8:423–438.

Beckett, C.T.S., Smith, J.C., Ciancio, D. & Augarde C.E. 2014. Observations on the effect of the addition of calcium chloride on the tensile strength of compacted soils under review for *Geoderma*, June 2014.

Brune, P.F., Ingraffea, A.R., Jackson, M.D. & Perucchio, R. 2013. The fracture toughness of an Imperial Roman mortar. *Engineering Fracture Mechanics*, 102: 65–76.

Bui, T.-T, Bui, Q.-B, Limam, A. & Maximilien, S. 2014. Failure of rammed earth walls: From observations to quantifications. *Construction and Building Materials*, 51:295–302.

Coghlan, S. 2014. *Bonding in fibre-reinforced earthen construction materials*. MEng Dissertation, Durham University, UK.

Corbin, A. & Augarde, C.E. 2014. Fracture Energy of Stabilised Rammed Earth. *Procedia Materials Science* 3: 1675–1680.

Gallipoli, D., Bruno, A.D., Perlot, C. & Salmon, N. 2014. Raw earth construction: is there a role for unsaturated soil mechanics? In *Unsaturated Soils: Research & Applications*, Khalili et al. (Eds), 55–62 CRC Press.

Gelard, D., Fontaine, L., Maximilien, S., Olagnon, C., Laurent, J-P., Houben, H. & Van Damme, H. 2007. When physics revisit earth construction: Recent advances in the understanding of the cohesion mechanisms of earthen materials in Proc. Int. Symp. Earth. Struct., Bangalore, 22–24 August 2007.

Hall, M. & Djerbib, Y. 2004. Rammed earth sample production: context, recommendations and consistency. *Construction and Building Materials* 18:281–286.

Houben, H. & Guillard, H. 1994. *Earth construction a comprehensive guide.* London: ITDG Publishing.

Howard, T. 2007. *Shear strength of rammed earth: investigation of strength differences through earth layers.* MEng Dissertation, Durham University, UK.

Jaquin, P.A. 2008. *Analysis of historic rammed earth construction,* PhD Thesis, Durham University, UK.

Jaquin, P.A., Augarde, C.E. & Gerrard, C.M. 2008. A chronological description of the spatial development of rammed earth techniques. *International Journal of Architectural Heritage: Conservation, Analysis, and Restoration,* 2(4):377–400.

Jaquin, P.A., Augarde, C.E., Gallipoli, D. & Toll, D.G. 2009. The strength of unstabilised rammed earth materials. *Géotechnique,* 59 (5):487–490.

Jaquin, P.A. & Augarde, C.E. 2012. *Earth building: history, science and conservation.* Bracknell: IHS BRE Press.

Maniatidis, V. & Walker, P. 2003. *A review of rammed earth construction.* University of Bath, Bath.

Nowamooz, H. & Chazallon, C. 2011. Finite element modelling of a rammed earth wall. *Construction and Building Materials,* 4: 2112–2121.

Readle, D. 2013. *The bonding of fibrous materials in rammed earth.* MEng Dissertation, Durham University, UK.

Smith, J.C. & Augarde, C.E. 2013. A new classification for soil mixtures with application to earthen construction. ECS Technical Report 2013/04 (www.dur.ac.uk/resources/ecs/research/technical_reports/SMCTechnicalPaper.pdf).

Smith, J.C. & Augarde, C.E. 2013a. Optimum water content tests for earthen construction materials. *ICE Proceedings: Construction Materials* 167(2): 114–123.

Smith, J.C. & Augarde, C.E. 2014. XRCT scanning of unsaturated soil: microstructure at different scales? in *Geomechanics from Micro to Macro,* Soga et al. (Eds), 1137–1142.

Smith, J.C., Augarde, C.E. & Beckett, C.T.S. 2014. The use of XRCT to investigate highly unsaturated soil mixtures. In Khalili et al. (eds), *Unsaturated Soils: Research and Applications,* 719–725.

Standards New Zealand 1998. *NZS 4297 Engineering design of earth buildings.* Wellington: Standards New Zealand.

Walker, P., Keable, R., Martin, J. & Maniatidis, V. 2005. *Rammed earth design and construction guidelines.* Watford: BRE Bookshop.

Zornberg, J.G. 2002. Discrete framework for limit equilibrium analysis of fibre-reinforced soil. *Géotechnique* 52:593–604.

Rammed Earth Construction – Ciancio & Beckett (Eds)
© 2015 Taylor & Francis Group, London, ISBN 978-1-138-02770-1

Rescuing a sustainable heritage: Prospects for traditional rammed earth housing in China today and tomorrow

R. Hu & J. Liu
School of Architecture, Xi'an University of Architecture and Technology, P.R. China

ABSTRACT: The paper gives a brief review of the current state of Rammed Earth (RE) housing in China. The review comprises the study of typical patterns of rammed earth houses in different areas, classification of historic and recent rammed earth houses, and an analysis of the changing nature and relationship with traditional rammed earth houses brought about by rapid urbanization. This review is based on a field study in villages in Shaanxi province, Sichuan province, Ningxia province, Yunnan province and Tibet. Two practical rammed earth housing projects are also presented in the paper. Aspects of sustainable development of rammed earth housing in China are discussed including building norms and the impact of occupants seeking a better living standard.

Keywords: Traditional rammed earth housing; west China; sustainable development

1 INTRODUCTION

The history of rammed construction techniques in China can be traced back more than 6000 years according to archeological findings in ruins of human habitats of The New Stone Age. The main portion of The Great Wall was built with rammed earth. The diversity of vernacular rammed earth architecture in China is a durable record of the history and culture of China. Rammed earth is considered a sustainable building material which is suitable for low cost housing. Unfortunately, the quality of most of the rammed earth houses in remote villages in China is poor. In the last 30 years in China, damages to earth dwellings including adobe and rammed earth were very serious and caused heavy casualties. More and more traditional forms of construction for houses are being replaced with new buildings that incorporate fired bricks and concrete. Knowledge of ancient rammed earth construction techniques are gradually being lost and face challenges in modern China.

2 THE TRADITIONAL RE HOUSE

Rammed earth construction techniques originated in the Central Plains of China and spread through migration. The diversity of climate, topography, historical background, nationalities and cultures created varied traditional rammed earth architectures. Combined with natural materials like wood and rock, the traditional RE dwellings are in harmony with the nature as if 'grown from the ground'.

2.1 Structure

Traditional RE structures in China can be classified into three types according to the bearing system: bearing-wall system, bearing-framework system and soil-wood composite bearing system. The position of the wood framework in the relation to the earth wall can be: inside the wall, on the inner side or the outer side. In the soil-wood composite bearing system both the earth wall and framework or columns of timber are load-bearing, for example, the Tibet watch-tower RE dwelling showing in Figure 1. Besides those most common structure types, RE is also used in cave houses. Figure 2 shows the adobe cave house in Shaanxi province, the piers, abutment and the upper part of the roof above the adobe arch are all RE.

Figure 1. Tibet watchtower style RE house (left) & schematic showing the structure of its living room (right).

Figure 2. Cave house.

2.2 Form

The form of the vernacular architecture is influenced by many factors in both the natural and the human environment. Climate is one of the important factors influencing the form of traditional dwellings. Besides climate, locally available building materials, lifestyle, custom, history, belief, aesthetic view point can also influence the form. Because of the complexity, here we only focus on the relationship of the traditional RE house form and the climate.

The traditional Chinese dwellings together with settlement design embody the traditional Chinese philosophy that 'human beings should be in harmony with nature'. The form of the traditional RE dwelling reflects the simple ecological wisdom which is consistent with today's basic concepts of passive design.

Most of the traditional RE dwellings in China are distributed in the western part of the nation. The northwest part of China is located in cold and severely cold climate zones. The general characteristics of the climate in this area are: long and cold winters, dry, high solar radiation, big diurnal temperature variation, windy and sandy weather appears during winter and spring. The main point of the houses from this area is to get more sunlight in winter, keep warm and avoid wind. The form of the RE traditional houses in this area is usually compact and enclosed. The general characteristics of the RE houses in northwest China are: facing south, compact form, low storey height about 2.2 m to 2.4 m, a compact and enclosed courtyard, thick wall, with thickness around 500 mm to 1000 mm, thick mud flat roof, bigger openings on the walls facing south whereas no windows or small windows in the walls in other directions.

The compact form and low storey height can reduce the heat loss from the exterior surface. The south facing windows let the winter sun in to increase the direct heat gain (northern hemisphere). Thicker earth walls have higher thermal mass and higher thermal resistance to keep cool in summer and warm in winter. No windows or small windows in non-southerly directions can protect from the cold northwest wind in winter and reduce heat loss. The flat roof also reduces the surface exposure to the wind. For dwellings with a courtyard such as the Zhuangguo house in Qinghai the courtyard is enclosed without windows on the exterior. Heavy RE walls protect from wind and maintain warmth.

In west China, the general trend of climate condition from the north to south, and west to east is: winters become shorter and less cold, warmer summers, less diurnal temperature variation, more precipitation and more humidity. In southwest China the form of the traditional RE house has developed to protect the earth wall from the rain, to shade sun for a cooler summer night and to get natural ventilation to both cool and remove the excessive moisture. The exterior form of the traditional RE house in this region becomes 'lighter' and 'more open'. Generally the thickness of the rammed earth wall is 350 mm to 450 mm based on the field study in Shaanxi, Sichuan and Yunnan. In many houses the facade RE wall is replaced by light material such like wood plank which is easier than traditional RE construction to include large windows and doors for natural ventilation.

The roof of the traditional RE house exhibits the general tendency to adapt to regional precipitation differences. In the area where the average annual precipitation is under 300 mm most of the traditional RE houses have a flat mud roof. In the area of average annual precipitation over 300 mm, the sloped roof becomes popular. Consistent with the average annual precipitation increasing, the degree of slope becomes bigger and length of the overhanging eave increases also to shed water and to get more shade. Most of the traditional RE houses have a two-sloped roof with the sloping degree in the range of 20° to 45° in the area where the average annual precipitation is over 500 mm. On the surface of sloped roofs, grey tiles are popularly used while wood shingles or slates are used in some mountain areas. The two-sloped roof creates an attic which is usually used for storage. In areas with a hot&humid summer and cold winter, windows on the gables are opened during summer to allow natural ventilation to both cool and remove excess moisture whereas windows on the gables are closed during winter to create a buffering space to keep the lower space warm.

Hakka RE dwellings are very unique for their gigantic volume, distributed in south China where the climate is warm and there is no winter. The extremely thick exterior wall and enclosed form were designed mainly for defense instead of for climate conditions.

3 THE RE HOUSE TODAY

RE dwellings in China today can be classified into three groups: 1) architectural culture relic; 2) non-engineered dwellings still being occupied; 3) engineered dwelling (new RE dwelling).

The first group is a traditional RE dwelling which has very high historical value and cultural value. For example, a traditional Hakka RE dwelling is on the list of World Culture Heritage Sites. The RE dwellings in this group are preserved or restored.

Most of the RE dwellings in China belong to the second group which are mostly distributed in remote poor rural areas in west China. The RE dwellings are occupant-self-built without architects and engineers.

The third group is composed of RE dwellings built with technical support from architects and engineers. In the last 20 years, more and more researchers, architects and engineers have started focusing on how to develop the traditional RE dwellings. However, the number of new RE dwellings is very limited.

Most of the non-engineered RE dwellings which are still occupied exhibit low quality with regard to structural safety and indoor environment. The traditional RE construction skills were accumulated and passed down from generation to generation by craftsmen mostly orally: such skills as addition and reinforcing to increase the strength of the wall, how to protect the earth wall from rain or moisture, and to increase the integration of the bearing structure. However, the traditional construction experiences and skills are disappearing causing the very low construction quality. Additionally, the old traditional RE dwelling pattern is not adaptable to modern life today. In rural areas, in pursuit of better living conditions, the occupants built their RE houses with bigger windows, higher and more storys, and bigger room sizes compared with the traditional RE house. Unfortunately without sufficient technical knowledge those changes make the houses more vulnerable in earthquakes.

General problems with non-engineered RE dwellings are discussed below.

3.1 Safety

West China is a very active earthquake region suffering from large earthquakes. Since 1920 about 90% earthquakes with a magnitude greater than 7.0 occurred in China were in west China. The low quality of the non-engineered RE houses can lead to deadly consequences during moderate to severe earthquakes.

The old traditional RE house is usually a compact box-type form with frequent cross-walls. This reflects the wisdom of principle earthquake-resistant construction for earth buildings. Wood bond beam, horizontal and vertical reinforcing made of timber, bamboo or natural fibers were popularly used in traditional RE construction. However, with growing population, declining wood resources and neglecting potential earthquake risk,

in many non-engineered RE construction those traditional methods to strengthen the wall and to increase the structural integrity are ignored. Typical seismic damage pattern and the problems of the un-engineered RE structures in rural area in China are broadly summarized in the following:

1. Overturn of the wall (Hu and Dong, 2010). The post-earthquake field survey shows the overturn of the wall from out-of-plane is one of the typical seismic damages for rammed earth walls. The main reasons include: poor connection between longitudinal wall and cross-wall which can cause the separation on the corner; low tension strength of the earth wall which causes severe vertical crack on the wall plane under bending moment. Additionally, to get more space many RE houses have very tall free-standing walls with a height-thickness ratio over 12 which also lead to the overturn of the wall.

2. Roof collapse (Hu and Liu, 2009). The earth roof built in the traditional way in cold areas is usually very heavy. It helps to insulate the building. However, during strong earthquakes, due to its heavy weight, the structure develops high levels of seismic force which the bearing structures are unable to resist and it experiences sudden failure.

3. Wall damage by shear force. Without vertical reinforcing, the interface between the lifts (RE block) is critical under lateral load which can lead to horizontal cracking.

4. Local damage of the earth wall supporting the ends of beams. The beams or rafters carrying the roof load are placed directly on the earth wall; local cracks in the wall appear even in the static state which leads to very severe potential damage during an earthquake.

5. Large openings in RE wall reducing the wall strength (Hu and Liu, 2009). To get more sunlight larger sized southern facing windows are set in the south facing earth wall. The total length of wall opening is often more than 50% of the wall length. Large windows reduce the strength of the wall facing south and also brings unclosed structural plane which may lead to twist pattern of earthquake damage.

6. Wall damage by moisture or uneven settling. The RE wall is damaged by moisture wicking up the wall for lack of a moisture-proof layer between the ground and the RE wall. Unstable base causes cracking due to the uneven settling.

In china there is no specific design code for RE construction. In The National Code for Seismic Design of Buildings (GB5011-2010) and Seismic Technical Specification for Buildings Construction in Town and Village (JGJ161-2008), only very general recommendations for both adobe and RE construction are mentioned which is inadequate for engineers to use as the basis for RE structural design.

Soil selection, particle distribution and additives are random in most non-engineered construction. With no quality supervision during and after the construction the safety of the house is hard to guarantee.

3.2 Thermal comfort

RE wall is considered to be a good thermal mass to keep a relatively stable indoor comfort with appropriate design parameters. However, the thermal environment during winter time in most RE houses in China is not satisfying. According to our field study in rural areas in west China more than 70% of RE occupants are satisfied with the indoor thermal environment during summer whereas more than 80% are not satisfied with the thermal environment during winter. Based on the test results in villages in Shaanxi, Sichuan and Yunnan province the average room temperature in RE houses is in the range of 0°C degree centigrade to 4°C degree centigrade with the average outside temperature of –2°C degree centigrade to –3°C (Hu et al., 2010; Hu et al., 2010a; Zhou, 2005). The acceptable indoor temperature for 80% occupants in rural area with cold winters ranges from 9°C to 15.8°C during winter, and the winter thermal neutral temperature is 11.6°C (Yang, 2010). Apparently, the thermal environment in the RE houses during winter are very poor and far below the occupants' comfort need. There are typically only one or two rooms using very inefficient stoves for heating intermittently and most rooms have no heater at all. One of the main reasons for the low indoor temperature is the cold air infiltration due to the lack of airtightness in most rural RE houses, which can also influence the thermal performance of RE walls. It also shows using only the exterior RE wall is not enough to create a comfortable indoor environment in winter in west China.

Figure 3 shows the test result of a RE house without heating in Qinmao village, Shaaxi in a typical winter's day. Room A is facing south and Room B is facing north. The average temperature of Room A is about 4°C higher than Room B indicating the solar heat gain is very important for RE wall's thermal performance during winter. Even with a sun-facing orientation and sun-facing glazed windows the temperatures of Room A are lower than 5°C degree centigrade indicating space heating and insulation to increase the thermal resistance of the RE wall are both needed.

3.3 Degrading of traditional RE dwelling

Because of the poor quality of the non-engineered rammed houses most villagers prefer to build fired brick masonry houses when the economic condition allows. Besides the low quality of both structure and winter thermal environment mentioned above, the inner space of the traditional RE house is not appropriate for modern lifestyles. Additionally, most of the non-engineered RE houses have the problem of poor daylighting and ventilation.

Figure 4 shows the decrease of RE construction in Qinmao village in Shaanxi province. Among the houses built during 2007 to 2014 only 10% of the houses used RE construction. Compared with brick masonry construction, traditional RE construction is cheaper on building material but it consumes more labor and time. With more and more able-bodied males leaving villages to work in urban areas, the labor cost in villages has risen and it is becoming hard to find builders who have RE construction experience.

For lack of knowledge about industrial building materials, the brick masonry houses built in most rural areas in China have both poor earthquake resistance capacity and thermal environment. The living conditions have not improved with transition to brick construction, or in many cases have become worse. Questionnaire results in villages in Shannxi, Ningxia, Sichuan and Yunnan shows more than 80% villagers prefer to build and live in a brick house instead of an RE house. The less than 20% who like to live in RE houses are mostly elderly. If the population all chose to live in fired brick structures, one outcome would be an increase in energy use, along with higher pollutant levels. With more fired brick houses built farm land would be

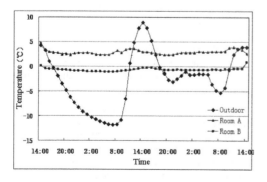

Figure 3. Room air temperatures and outdoor air temperature for a RE house in Shaanxi.

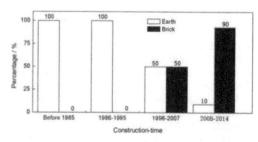

Figure 4. Decrease of RE construction in Qinmao village in Shaanxi province.

reduced and these modern developments would have a detrimental effect on the natural ecosystem. Furthermore, a significant vernacular culture embodied in traditional RE dwellings would disappear. We have already seen the loss of indigenous environments in the rapid urbanization of China's eastern cities. How to inherit and develop the traditional RE construction is a big challenge in China.

4 ENGINEERED RE DWELLING

The traditional RE houses have their own deficiencies which cannot meet modern living requirement, thus, need to be updated. Researchers in China have been working on RE practical projects to explore an appropriate approach of sustainable RE houses especially in rural areas. In those projects the essence of the sustainable development of traditional RE houses in China is applying modern science and environmental knowledge to the rural house design while respecting the wisdom of traditional techniques with consideration of local culture to meet the need of the occupants. Two representative projects are briefly introduced here.

4.1 *Project 1 (Liu et al., 2010)*

The project is located in a rural region in Yong Ren County, Yunnan Province. The project was to build demonstration RE houses for the minority occupants whose traditional homes were made of earth and wood. The field survey and research commenced in 2002 in which both the merits and defects of the traditional houses were evaluated. The culture of the local minority and their typical lifestyle patterns were obtained from the research. A questionnaire was used to understand basic living requirements of the local people. The properties of the soil from the new settlement were laboratory tested, with different cost-efficient additives. The appropriate additives and their proportions were derived to get the permissible compressive strength of the earth wall. Static and dynamic structural experiments were conducted to check the earthquake-resistant capacity of the earth walls. Traditional housing techniques and principles have been applied and upgraded to make the buildings seismically stable. Natural ventilation and solar heating are organically combined in new designs after related simulation analysis. Local people have been involved in the design and construction processes. By 2006 more than 400,000 square meters of new earth housing had already been constructed according to the design. The research group compared the thermal environment, daylighting factors of old and new RE dwellings by site measurements. The results show that the indoor thermal features and daylighting factor of new RE house is much better than old the one's. Feedback from the villagers showed that most of them were satisfied with the new earth houses. Figure 5 and Figure 6 show the images of the wall during construction and the demonstrate new RE house.

4.2 *Project 2 (Zhou et al., 2012)*

This project is about post-earthquake construction in Maanqiao village in Gansu province after 2008 Wenchuan earthquake. In the project traditional RE formwork and ramming hammers are improved which deliver a better construction quality result. Traditional methods to strengthen the RE wall which have been proved through tests are used in the construction. As shown in Figure 7, bamboo is used for horizontal reinforcement and stick is set to lock into the lifts. The structural integrity is

Figure 5. Rammed earth walls during construction.

Figure 6. Demonstrate RE house in Yunnan.

Figure 7. Traditional reinforcing in RE wall[11].

Figure 8.   Demonstrate RE house in Gansu[10].

increased due to the wood bonding beams and wood posts set in each corner (Yang, 2010). In later housing construction with international collaboration, a mechanical rammed earth method is introduced and concrete bonding beams are used. Local laborers were trained during the project which help spread the RE construction techniques and also increased the construction quality. This project became a demonstration post-earthquake RE house project.

5 CONCLUSION

In summary, to upgrade the traditional RE construction is representative of a larger problem currently facing Chinese architecture in general. The root of Chinese culture is in rural areas where more than half of the population is living. Vernacular architectural culture is a significant part of our history and culture. With China's rapid economic development, urbanization and globalization, vernacular architecture including traditional RE dwellings is being replaced by conventional structures. In the rapidly developing economy, local populations wish to improve their rural living condition at the least cost, but few have any knowledge of energy and resource saving, regional cultural heritage and sustainable developing. The goal of upgrading traditional RE construction should put improving living conditions as the key point. The opinion of occupants about the 'poor quality' of RE dwellings can only be changed by quality RE demonstration projects which stay true to regional traditions and yet offer all the amenities of modernization. The application of modern science and environmental knowledge to the traditional house designs can make huge improvements in the quality of the RE homes and comfort levels and win people back to the benefits of RE dwellings. It is believed that traditional wisdom and lore in buildings may still offer wisely managed, economically effective and culturally appropriate solutions to the world's housing needs (Oliver, 1997). Considering the diversity of cultures in different areas in China, local RE construction standards with local culture considerations are quiet necessary. The seismic rural area where most of buildings are non-engineered is the place that needs the architectural and engineering supports most. Training and having the local construction team is a feasible way to guide a sustainable construction of RE houses.

ACKNOWLEDGEMENT

The research is supported by National Natural Science Foundation of China (Project number: 51378410 & 51221865).

REFERENCES

Hu Rongrong, Dong Yujiang. Shake table test on rammed earth wall panels. Advanced Materials Research, 2010, Vol (133–134):133–134. In English.

Hu Rongrong, Li Wanpeng, He Wenfang, Liu Jiaping. Building space form and thermal environment study of rural houses in south Shanxi mountain area. International Conference on Building Environment Science & Technology 2010: 162–166. In Chinese.

Hu Rongrong, Liu Jiaping. Study on the evolution of the vernacular houses in the rural areas in Tibet. [J]. Journal Of Xi'an University Of Architecture & Technology, 2009, 41(03):380–384. In Chinese.

Hu Rongrong, Wang Peng, Yang Liu, Liu Jiaping. The analysis of the winter thermal property of brick-concrete houses and soil dwelling in Guanzhong region. International Conference on Building Environment Science & Technology 2010a: 422–429. In Chinese.

Liu Jiaping, Hu Rongrong, Wang Runshan, Yang Liu. Regeneration of vernacular architecture: new rammed earth houses on the upper reaches of the Yangtze River, Frontiers of Energy and Power Engineering in China, 2010, 4(01):93–99. In English.

Ministry of Housing and Urban-Rural Development of the People's Republic of China, GB5011-2010 Code for Seismic Design of Buildings, China Architecture & Building Press, 2010. In Chinese.

Ministry of Housing and Urban-Rural Development of the People's Republic of China, JGJ161-2008 Seismic technical specification for buildings construction in town and village, China Architecture & Building Press, 2008. In Chinese.

Oliver P. (1997), "Encyclopedia of Vernacular Architecture of the World," Volume1, Cambridge University Press, Cambridge, 134.

Yang Hua. Seismic Test of Rammed Earth Wall and Reconstruction of Demonstrate Residence After Panzhihua Earthquake. Dissertation, Xi'an University of Architecture & Technology 2010. In Chinese.

Yang Qian, Study on the Indoor Thermal Comfort in the Cold Zone. Dissertation, Xi'an University of Architecture & Technology 2010. In Chinese.

Zhou Tiegang, Yang Liping, Mu Jun. Research & Application of Contemporary Rammed Earth Rural House. Construction Science & Technology. 2012, Vol 23: 58–61. In Chinese.

Zhou Wei. Study on the Analysis of Architectural Space and the Regeneration of Traditional Dwellings. Dissertation, Xi'an University of Architecture & Technology 2005. In Chinese.

Delegates' papers

Rammed Earth Construction – Ciancio & Beckett (Eds)
© *2015 Taylor & Francis Group, London, ISBN 978-1-138-02770-1*

Improved thermal capacity of rammed earth by the inclusion of natural fibres

G. Barbeta Solà & F.X. Massó Ros
Department of Architecture and Construction Engineering, University of Girona, Girona, Spain

ABSTRACT: In temperate climates, rammed earth walls provide great thermal inertia and adequate insulation, depending on soil type, density and thickness. However, when designing walls of the colder facades of a zero energy building, or an almost entirely passive building, a higher thermal insulation rate is needed for the walls in order to optimise efficiency. Although additional construction solutions are possible, such as the insertion of an insulation layer, these may be difficult to implement when the rammed earth finish is preferred. This paper proposes an alternative solution, involving the incorporation of natural and recycled fibres in the earth block composition. Preferably, local materials should be used. Brown fibres, certain expansive clays (vermiculite, perlite, arlite) and natural cork shavings of varying grain sizes are some of the possible materials. This study reveals a significant improvement in thermal conductivity in all cases—up to 60% in the most optimal compositions.

1 INTRODUCTION

At times, the thermal inadequacy of rammed earth buildings has become evident when attempting to comply with the energy saving requirements governing the Basic HE Document of the Spanish Technical Building Code (CTE DB HE, for its initials in Spanish). In designs having walls that consist mainly of rammed earth without cladding, insulation indexes are typically in the order of 0.6 at 1.2 W/mK (Bauluz & Bárcena, 1991) for standard thicknesses of between 30 and 60 cm, also depending on the obtained density, amount of clay and soil type.

Therefore, when comparing standardized conductivity indexes for the core materials of conventional constructions and those made of earth, the rammed earth type lies in a middle position.

In order to comply with the maximum thermal transmittance U, which according to climatic areas is 0.74 at 1.22 in W/m2 K for walls and interior partitions of the thermal envelope, the walls need to measure approximately one meter in thickness.

Therefore, the inclusion of internal or lightening insulators in the composition of rammed earth walls is a practice that, in accordance with other studies, has been found to generate significant improvements in this area, despite requiring special monitoring of the loss of other attributes, including compressive strength.

2 BACKGROUND

2.1 *The use of black cork shavings*

In 1994 the authors used cork as a stabilizer and additive to improve particle size and thermal quality. Initial tests were conducted with "Licorela" type earth and shale with additions of 20–30% per volume of cork and arlite, a lightweight aggregate derived from expanded clay based on thermal treatment and autoclaving. This work has continued with two current rammed earth building projects in Barcelona using the manufacturing process of a structural prefabricated piece, armed with bamboo rots (Figure 1). Crushed black cork was used from the remains of the cork panel manufacturing process by pressure and temperature. The obtained particle size was in the order of 12/25. This paper presents the sample results as nomenclatures C and CC.

2.2 *Arlite or expanded clay*

Arlite (also known as ripiolite, expanded clay or leca) is a lightweight ceramic aggregate.

Figure 1. Rammed earth building in Collserola, Barcelona. Above: Prefabricated cement-cork stabilised earth ground floor slab undergoing laboratory testing.

The S-low modular building system consisting of a wood frame and rammed earth, developed in Barcelona by GICITED (Gonzalez Sanchez, 2013), uses walls stabilized on a 10 cm thick central strip with 12.2% arlite in volume. Thus, an average dry density of 2.07 g/cm³ is obtained, equivalent to proctor compaction of 73.75%, improving insulation by 25%.

2.3 Using perlite and wood fibres in a bioconstructive and bioclimatic public school

Another earth construction, in Santa Eulalia de Ronçana (Barcelona), was the winning proposal (out of twelve offers) in a request for tenders. Additionally, the project was recognized by the European Awards and has received two awards for sustainability and environmental quality: the Ecoviure and the Endesa Awards 2010. As an ecological gesture, earth from the site was used. The use of earth from the excavation site to build walls is a way to clearly minimize construction impact. Although linear shrinkage is a major handicap, it also provides water protection.

Table 1. Comparative values of rammed earth insulation.

Material/Reference	Density	Conductivity λ
Rammed earth, sun-dried brick, Compressed Earth Block (CEB) (*Eduardo Torroja Construction Sciences Institute, CEPCO & AICIA, 2010*)	1770–2000 kg/m³	1,10 W/mk
Rammed earth (*Bauluz & Bárcena, 1991*)	1400 kg/m³	0,60 W/mk
	1600 kg/m³	0,80 W/mk
	1800 kg/m³	1,00 W/mk
	1900 kg/m³	1,20 W/mk
	2000 kg/m³	1,60 W/mk
CEB (*Dominguez, 1998*)	1700 kg/m³	0,81 W/mk
CEB recycled C UdG BTC UNE 92 (*Alvarez Vazquez & Potrony Serret, 2001*)	1960 kg/m³	0,41 W/mk
CEB recycled 1 UdG BTC UNE 92 (*Alvarez Vazquez & Potrony Serret, 2001*)	1710 kg/m³	0,54 W/mk
Sun-dried brick (*Dominguez, 1998*)	1200 kg/m³	0,46 W/mk
BTC Bioterre (*Bioterre technical specifications*)	1790 kg/m³	0,48 W/mk
Cannabric (*Cannabric technical specification*)	1171 kg/m³	0,19 W/mk
Reinforced concrete	2300–2500 kg/m³	2,3 W/mk
Mass concrete in stiu	2000–2300 kg/m³	1,65 W/mk
Baked clay for masonry units (*Eduardo Torroja Construction Sciences Institute, CEPCO & AICIA, 2010*)	1700–2000 kg/m³	0,59/0,74 W/mk

3 HYPOTHESIS FOR IMPROVEMENT

3.1 *Type of soil utilized*

The soil SC is composed of arkosic sandstone, with clay from the Pliocene era and granitoid meta-sediments from the Palaeozoic era, including shale, gneiss and quartz gravels. The DRX has also revealed minority components such as albite and sanidine. Their presence occurs in magnitudes of 5%, quite consistent with the behaviour observed in sediment measurements. The primary clay present in the C and CC soils is chlorite, although there is a higher content of calcium in the second. The method of clay type determination used in this case was DTA/DTG, differential thermo gravimetric analysis. Figures 2 and 3 show the different peaks where major mass loss occurs.

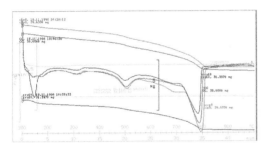

Figure 2. DTA-thermo gravimetric analysis. CC samples. Chlorite content is evident.

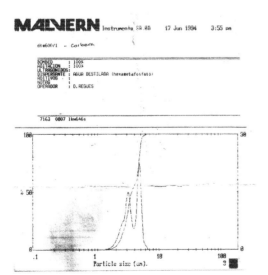

Figure 3. Sedimentometry with hexametaphosphate. The large particle size of the C sample confirms Chlorite presence.

3.2 *Physical and chemical analysis*

Particle size distribution of the unbalanced SC and LL particles is to be in accordance with standards established for the execution of the rammed earth technique. LL soil must contain 20–30% of a select recycled sand in order to compensate for the excess of silt, as revealed in the graph. SC particle size distribution consists of a much higher percentage of <2 mm particles, contrasting with a very low content of medium and large particles (fine gravel, gravel and stones). According to the USDA ternary graph classification, it is clay loam soil.

The SC plasticity index of particles <0.32 mm (fine sands, silts and clays) is approximately 12.7%. The percentage of calcium carbonate is quite low or non-existent, according to the qualitative determination of reaction generated by the hydrochloric acid attack.

The presence of organic matter has been quantified at slightly below 1% in all soils. Linear shrinkage value is 3.5% in SC and 4.5% in LL plane soil.

3.3 *Characteristics of rammed earth stabilizers*

Natural cork: Raw material typical of the Mediterranean region, quite abundant in the Iberian Peninsula and in much of the overall Girona province. It is a natural, sustainable material that fosters regional forest management.

For this project, four types of natural cork shavings were used (by-product 0.5–2, by-product 2–3, Broken and Powder or Earth).

Arlite: Expanded clay aggregate is a lightweight ceramic shell with honeycomb core produced by firing natural clay to temperatures of 1100–1200 °C in a rotating kiln. The pellets are rounded in shape and fall from the kiln in a grade of approximately 0–32 mm with an average dry bulk density of approximately 350 kg/m³. It's not susceptible to chemical attack, rot or frost and has a long life span.

Triturated almond Shell: This is a by-product of the processing of nuts. Granulometry is between 0.95–9 cm and density is 1.1 g/cm³.

Olive stone triturated and cleaned: This is a by-product produced in the industrial production of olive oil. Granulometry is between 0.9–3 cm.

4 LABORATORY TESTS

4.1 *Dosages*

Five different compositions were made with varying percentages and types of each natural fibre, production water and cementitious stabilizers (Table 2). Different methods of dynamic and static compression of 2 MPa were used. Samples were cured for 28 days.

Table 2. Dosages and results of the linear shrinkage test.

Composition nomenclature	Type of fibers	Fibers percentage	Chemical stabilizer percentage	Linear shrinkage percentage
SC0 (*R1*)	–	–	–	3.66%
SC8P (R2.1)	–	–	8% CP	1,50%
SC8P30S (*R3.1*)	Cork shavings (by-product 0,5–2)	30,00%	8% CP	1,00%
SC8P20S (*R10*)	Cork shavings ("Powder" or "Earth")	20,00%	8% CP	1,08%
SC8P35S (*R19*)	Cork shavings (By-p. 0,5–2 + By-p. 2–3 + "Broken")	35%*	8% CP	0,83%
LL0 (*Comp. 1*)	–	–	–	4.46%
LL5P (*Comp. 3*)	–	–**	5% CP	0,98%
LL5P30CA (*Comp. 5*)	Almond shell	30,00%	5% CP	1,93%
LL5P20CA (*Comp. 6*)	Almond shell	20,00%	5% CP	1,35%
LL5P30HO (*Comp. 7*)	Olive stone	30,00%	5% CP	1,20%
CC0 (*CC*)	–	–	–	–
CC8P40S (*CC*)	Cork shavings (Black cork shavings)	40,00%	8% CP	–
CC8P40A (*CC*)	Arlite (little gradding arlite)	15,00%	8% CP	–
C0 (*C*)	–	–	–	–
C8P40A (*C*)	Arlite (Big gradding arlite)	40,00%	8% CP	–
GU8P40F (*GU*)	Wood fibers	40,00%	8% CP	–

CP: Portland Cement CEM II.
*By-product 0,5–2 (12,5%) + By-product 2–3 (12,5%) + "Broken" (10%).
**Optimal physical stabilization with earth particles.

Rejected
Accepted

4.2 Densities

The main objective of including natural fibres is to attain a material having increased insulation properties. Also, weight is significantly reduced.

4.3 Volume stability—linear shrinkage

Local soil that was not subjected to stabilising processes had a very high linear shrinkage level (4.5–3.5%), which, due to the rammed earth construction method, compromised its usage. Thus, on a theoretical level, it was believed that the application of physical and chemical stabilizers in varying percentages would offer significant improvements in this respect. A linear shrinkage test was conducted for each of the sample compositions (Barbeta Solà, 2002), an acceptable limit was established at <20 mm/m or ≥1%.

4.4 Breakage under compression

Indeed, in all cases, results reveal considerable strength, except for the wood fibres. The worst of these have a high percentage exceeding 40%, specifically, in the arlite samples.

4.5 Thermal tests

A standard test was carried out using UNE standard 92 204:1995 and the tests conducted at the University of Saskatchewan (Hutcheon & WH, 1949) as a reference. In order to identify the thermal conduction coefficient of the studied material, the flow of heat transmitted from a hot side to a cold side was determined by measuring air temperatures and the surfaces on both sides of the sample, as well as the flow of heat from one chamber in a steady state.

Table 3. Levels obtained under compression.

Composition nomenclature	Density	Resistance capacity— standarized levels	Reducing percentage
SC0 (*R1*)	2080 kg/m³	5,40 N/mm²	–
SC8P (*R2.1*)	1830 kg/m³	6,29 N/mm²	–
SC8P30S (*R3.1*)	1540 kg/m³	3,91 N/mm²	38,00%*
SC8P20S (*R10*)	1720 kg/m³	4,08 N/mm²	35,00%*
SC8P35S (*R19*)	1490 kg/m³	3,98 N/mm²	37,00%*
LL5P (*Comp. 3*)	1970 kg/m³	2,76 N/mm²	–
LL5P30CA (*Comp. 5*)	1550 kg/m³	2,08 N/mm²	25,00%**
LL5P20CA (*Comp. 6*)	1700 kg/m³	2,04 N/mm²	26,00%**
LL5P30HO (*Comp. 7*)	1550 kg/m³	1,60 N/mm²	42,00%**

*Reducing percentage is calculated respect the resistance of test piece SCP8.
**Reducing percentage is calculated respect the resistance of test piece LLP5.

Composition nomenclature	Density	Resistance capacity— standardized levels	Reducing percentage
CC0 (*CC*)	2070 kg/m³	2,72 N/mm²	–
CC8P40S (*CC*)	1340 kg/m³	1,83 N/mm²	66,9%**
CC8P40A (*CC*)	1910 kg/m³	1,47 N/mm²	73,4%**
C0 (*C*)	2150 kg/m³	1,73 N/mm²	–
C8P40A (*C*)	1500 kg/m³	1,45 N/mm²	73,75%*
GU8P40F (*GU*)	1700 kg/m³	5,138 N/mm²	112%***

*Reducing percentage is calculated respect the resistance of test piece C8P; 4,32 N/mm².
**Reducing percentage is calculated respect the resistance of test piece CC8P; 5,54 N/mm².
***Increasing percentage is calculated respect GU8P; 2,45 N/mm².

5 CONCLUSIONS

Portland cement is the factor which, in this case, results in the greatest reduction in linear shrinkage—by 50%. The inclusion of different types, sizes and percentages of cork, arlite, perlite or wood fibres instead of only one results in slight improvements in volume stability, allowing for

Table 4. Thermal conductivity levels obtained.

Composition nomenclature	Density	Conducti- vity λ	Reducing percentage
SC0 (*R1*)	2080 kg/m³	1,35 W/mk	–
SC8P (*R2.1*)	1830 kg/m³	0,89 W/mk	–
SC8P30S (*R3.1*)	1540 kg/m³	0,45 W/mk	66,70%*
SC8P20S (*R10*)	1720 kg/m³	0,50 W/mk	63,00%*
SC8P35S (*R19*)	1490 kg/m³	0,56 W/mk	59,00%*
LL5P (*Comp. 3*)	1970 kg/m³	0,93 W/mk	–
LL5P30CA (*Comp. 5*)	1550 kg/m³	0,33 W/mk	64,00%**
LL5P20CA (*Comp. 6*)	1700 kg/m³	0,40 W/mk	57,00%**
LL5P30HO (*Comp. 7*)	1550 kg/m³	0,42 W/mk	55,00%**
CC0 (*CC*)	2070 kg/m³	0,50 W/mk	–
CC8P40S (*CC*)	1340 kg/m³	0,19 W/mk	62,00%***
CC8P40A (*CC*)	1910 kg/m³	0,25 W/mk	50,00%***
C0 (*C*)	2150 kg/m³	0,46 W/mk	–
C8P40A (*C*)	1500 kg/m³	0,38 W/mk	17,00%****
GU8P40F (*GU*)	1700 kg/m³	0,54 W/mk	51,00%*****

*Reducing percentage is calculated respect the conductivity result of test piece SC0.
**Reducing percentage is calculated respect the conductivity result of test piece LLP5.
***Reducing percentage is calculated respect the conductivity result of test piece CC0.
****Reducing percentage is calculated respect the conductivity result of test piece C0.
*****Reducing percentage is calculated respect the rammed earth standarized level of conductivity 1,10 W/mK.

compliance with the acceptance limit of ≥1% in all cases. The stabilization applied has proven effective in improving volume stability and compressive strength and thermal conductivity has also increased substantially. It demonstrates that the best percentage it's around 30% of similar fibbers to reduce more than 50% in thermal conductivity.

Using a homogeneous mixture in the overall rammed earth matter is a very attractive option, as improvement methods in different layers greatly complicate production, hindering the compression and pouring of the material.

REFERENCES

AEN/CTN 41 Construction/SC 10. 2005. *Standard UNE 41410 Compressed Earth Blocs for walls and partitions. Definitions, specifications and test methods*. Madrid: AENOR.

Barbeta Solà, G. 2002. *Doctorate thesis. Improvement of stabilised soil in the development of sustainable architecture towards the 21st Century*. Barcelona: Higher Technical School of Architecture.

Bauluz, G. & Bárcena, P. 1991. *Basic rules for designing and building with rammed earth*. Madrid: Spanish Ministry of Public Works and Urbanism.

Dominguez. M. 1998. *Thermal properties of adobe CSIC. Architecture of soil*. Madrid: Spanish Ministry of Public Works and Transport.

González Sanchez, B. 2013. *P.F.G Report on the thermal and mechanical properties of the S-Low modular construction method with a wood structure and rammed earth cladding*. Barcelona: Higher Technical School of Architecture.

Hutcheon, N.B and WH. Ball. 1949. *Thermal conductivity of Rammed Earth*. University of Saskatchewan: C. Engineering.

Rammed Earth Construction – Ciancio & Beckett (Eds)
© 2015 Taylor & Francis Group, London, ISBN 978-1-138-02770-1

Strengthening mechanisms in Cement-Stabilised Rammed Earth

C. Beckett & D. Ciancio
The University of Western Australia, Crawley, Australia

S. Manzi & M. Bignozzi
The University of Bologna, Bologna, Italy

ABSTRACT: There is currently little scientific understanding of stabilised Rammed Earth (RE) and the relationship between water-cement ratio and compressive strength. For traditional (unstabilised) RE materials, it is standard practice to compact the soil mix at its optimum water content to achieve maximum dry density and hence maximum strength. However, this may not also apply to Cement-Stabilised Rammed Earth (CSRE). A recent investigation (Beckett and Ciancio 2014) showed that CSRE samples stabilised with 5% cement and compacted at a water content lower than optimum performed better than samples compacted at optimum or higher. This seems to be in agreement with the well-known effect in concrete materials, according to which the lower the water-cement ratio, the stronger the cementitious products hence the higher the compressive strength. This paper investigates the effect of water cement ratio in CSRE samples. Results of an experimental programme are presented and used to discuss the appropriateness of the water-cement ratio for RE materials.

1 INTRODUCTION

Cement stabilisation is now commonplace in rammed earth (CSRE) construction. Although it has been acknowledged by several authors that there is a significant reduction in its environmental sustainability (e.g. Venkatarama Reddy and Prasanna Kumar 2010), the associated increase in material strength and durability is undeniable. However, what is less clear is how best to control the effects of cement stabilisation to achieve the maximum material improvement for the least cost, both environmental and financial.

Venkatarama Reddy and Prasana Kumar (2011a, 2011b) investigated several aspects of CSRE construction. For all of the materials tested, an increase in Unconfined Compressive Strength (UCS) was found with compaction water contents increasing from below to above the Optimum Water Content (OWC). Similarly, materials compacted at a given water content but to a range of dry densities (ρ_d) also showed increasing UCS with increasing ρ_d. Water content at testing was also examined; specimens dried at 50°C showed significantly higher UCS than similar specimens which had been dried and then submerged in water for 48 hours. This result is consistent with those found by Jaquin et al. 2009, and later Bui et al. 2014, for unstabilised RE, who demonstrated the strong link between suction present in RE's water phase and material strength. This suggests that suction phenomena also play a key role in CSRE.

Cement hydration mechanisms in CSRE were investigated by Beckett and Ciancio 2014. In that work, wrapped specimens were used to determine amounts of water used in cement hydration for specimens compacted above, at and below their OWC. Specimens with lower compaction water contents were found to have higher UCSs for all tested hydration times (between 1 and 28 days), contrary to findings in Venkatarama Reddy and Prasanna Kumar 2011a. Somewhat counter-intuitively, these specimens also had marginally lower dry densities than the other specimens tested and contained the least amount of hydrated cement. It was suggested that specimen strengths were therefore dependent on cement distribution and a transition between 'bridge' and 'matrix'-dominated cement regimes as compaction water contents increased.

It is clear from these works (and numerous others that cannot be covered here) that the interaction between cement, water and soil in CSRE is far more complicated than it has been credited with in the past. This paper presents an experimental investigation in which compaction water content, dry density and cement contents are closely controlled in order to more clearly discern the effects of each component on subsequent material strengths. Details of the experimental programme are given in the following section, after which results are presented and discussed in the light of those found by previous authors.

2 EXPERIMENTAL PROGRAMME

2.1 *Material*

An engineered soil, manufactured from controlled quantities of silt (Unimin Silica 200G), sand (an equal mix of Unimin SF and RC sand) and blue metal aggregate (max size 10 mm) was selected for testing. Clay was not added to the mix as it has been shown to interfere with cement hydration; although this would be more representative of the behaviour of stabilized natural soils, clay was omitted for improved experimental consistency. The particle size distribution for the final mix is shown in Figure 1. Cement contents of 5, 10 and 15% were selected for testing to represent typical stabiliser quantities used in RE construction and added to the dry soil mix. To improve consistency, specimens were manufactured from individual batches of each mix.

2.2 *Compaction water contents and specimen manufacture*

CSRE OWCs were determined using the Modified Proctor test as per AS 1289.5.2.1 (Standards Australia 2003). Water was added to dry soil and cement mixes in *a priori* known amounts and mixed for 5 minutes to ensure, as far as practicable, uniform water distributions throughout the material. Wetted mixes were compacted within 45 minutes of water addition.

Compaction curves for each mix are shown in Figure 2, where compaction water content, w, has been normalised via $w' = w'/owc$. OWCs for 5, 10 and 15% cement stabilised mixes were 7.9, 8.1 and 8.7% (by mass) respectively. Four values of w' were then chosen for specimen manufacture as shown in Figure 2: $w' = 0.76$ and $w' = 0.89$ (corresponding to measured datapoints for 5% cement content), selected to determine whether an optimum strength existed for materials compacted <OWC, as found in Beckett and Ciancio 2014; $w' = 1$ to investigate behaviour of specimens compacted to their maximum dry density ($\rho_{d,max}$); and $w' > 1$, corresponding to ρ_d values equal to those at $w' = 0.89$ for that mix. Interestingly, Figure 2 shows that the compaction curve for 10% cement falls below those of 5 and 15% for all but the highest values of w'. This is contrary to results found by Bryan 1988 and later by Venkatarama Reddy and Prasanna Kumar 2011a, who found either unchanging or increasing $\rho_{d,max}$ with increasing cement content and serves to highlight the variability inherent in earthen materials.

⌀100 mm, 200 mm tall UCS specimens were compacted in five equal layers of controlled mass and volume to ensure correct compacted densities. Once compacted, specimens were removed from the mould and cured under conditions of 94 ± 2% humidity, 21 ± 1°C for 28 days to ensure suction equilibration. Specimen UCS was then immediately determined by uniaxial crushing at a rate of 0.3 mm/min until failure, preventing re-equilibration to atmospheric conditions. Crushed material was oven-dried for 24 hours at 105°C to determine its free water content and ρ_d.

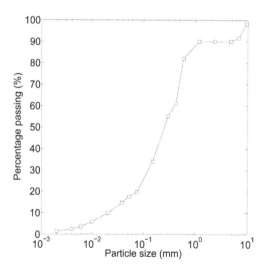

Figure 1. Particle grading curve for engineered soil mix.

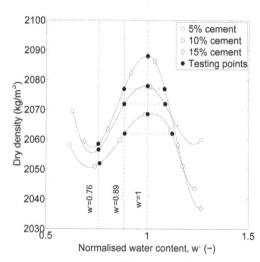

Figure 2. Normalised compaction curves for the three tested stabilizer contents.

3 EXPERIMENTAL RESULTS

Figure 3 shows results for specimen UCS against w', where trends through average UCS values for each cement content have been added for clarity. Figure 3 shows similar behaviour for all tested cement contents, in that maximum UCS values are, within the scatter of the data, largely obtained by materials compacted between $w' = 0.89$ and $w' = 1$. This is consistent with findings of Beckett and Ciancio 2014. Differences between maximum and minimum strengths in Figure 3 per given mix are also of similar magnitude to those found in that work (roughly 2 MPa). Relationships between w' and mix strengths are shown in greater detail in Figure 4, where specimen UCS has been normalised via UCS' = UCS/UCS$_{max}$ and where UCS$_{max}$ is the maximum average UCS found for that material.

Figure 3 shows that, with the exception of specimens manufactured at 5% cement content, specimens manufactured to the same ρ_d values above and below $w' = 1$ achieved roughly identical strengths, seemingly contradicting results found in Beckett and Ciancio (2014). This is investigated in more detail in Figure 5, which shows changes in specimen ρ_d between 0 (white markers) at 28 days (black markers) due to cement gel growth. Note that ρ_d values at compaction shown in Figure 5 are lower than the target values shown in Figure 2; this is due to the need to trim specimens once compacted to achieve a smooth testing surface.

For all values of w', Figure 5 shows significantly larger increases in ρ_d for specimens manufactured at higher cement contents; this is to be expected, due to the larger volume of cement gel

created. However, larger changes in ρ_d are seen for $w' > 1$ than for $w' = 0.89$, despite their similar densities. This is consistent with Beckett and Ciancio 2014 and suggests that greater amounts of hydrated cement were present in specimens compacted $w' > 1$ than for those at $w' = 0:89$. An exception to this is again seen for 5% cement specimens made at $w' > 1$. However, Figure 5 shows that compacted ρ_d values for 5% specimens manufactured at $w' = 0.89$ were lower than those compacted at $w' > 1$. Notably, the latter specimens achieved the lowest value of UCS of all tested specimens, despite their apparent 'advantage' of a

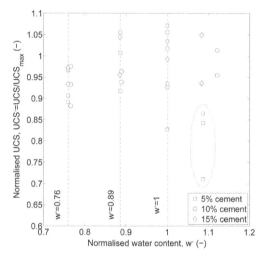

Figure 4. Normalised UCS results.

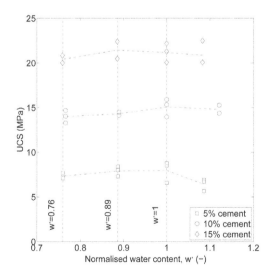

Figure 3. UCS results for specimens compacted at controlled values of w'.

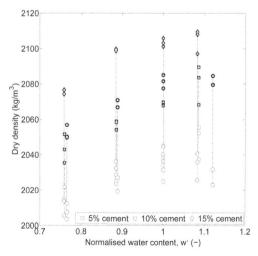

Figure 5. Evolution of specimen dry densities. White markers: ρ_d at compaction; black markers: ρ_d at 28 days. Arrows show transition between 0 and 28 days.

43

higher compacted density. It is suggested that, for $w' > 1$ 5% specimens, higher values of ρ_d resulted in a reduced cement gel interconnectivity, due to the reduced porosity, and hence lower strengths. If results corresponding to $w' > 1$ for 5% cement are discounted from Figure 4 (circled), a relatively consistent trend is seen between all cement contents, with strengths peaking as $w' \to 1$ and reducing thereafter, as identified above.

It is well known that the water-cement ratio (w/c) is a key factor in controlling the strength of concrete mixtures (Neville 2011). Given the similarity in their components, it has therefore been suggested that w/c plays a similarly important role in CSRE materials (Ciancio, Jaquin, & Walker 2013).

Specimen w/c against UCS are shown in Figure 6. Though results in Figure 6 across all tested materials suggest that a decrease in w/c results in an increase in strength, as is the case for concrete, the outcome that strength increases with cement content is largely trivial. Interestingly, however, results in Figure 6 show no apparent relation between w/c and UCS for constant cement contents, i.e. no increase in strength is observed when w/c reduces through a reduction in w'. This is illustrated in Figure 6 by results found for $w' = 0:76$ for 10% cement and $w' \geq 1$ for 15% cement; although the two materials achieved very different strengths, their w/c values are equal. A similar result was found by Fernandes et al. 2007 for compacted clay-sand-cement mixes. In that work, a strong trend between UCS and w/c was found

for mixtures compacted at or above $w' = 1$, with UCS rapidly decreasing for mixtures compacted below their OWC. It is therefore clear that microstructural phenomena regarding the distribution of the cement and soil aggregates play a key role in determining material strengths beyond a simple ratio between water and cement. An explanation might be that, in concrete, all of the water in the mixture (with the exception of that lost to evaporation) is available for cement hydration by virtue of its fluidic, saturated nature (Neville 2011). In CSRE, however, there is competition for water between the cement and the dry soil. Evidence for such a mechanism is suggested in Beckett and Ciancio 2014, however testing was not conducted over a range of cement contents. This is therefore a subject for further testing; microstructural investigations using SEM are ongoing.

4 CONCLUSIONS

This paper has presented experimental work investigating the role of compaction water content in controlling cementation mechanisms present in CSRE materials. Results showed similar trends in strength between specimens compacted to specific values of w' at different cement contents, demonstrating the strong role played by particle aggregation in strength development in CSRE. Material w/c were also investigated, showing that these alone are insufficient to describe cementation mechanisms in CSRE, suggestibly due to the different hydration environments between CSRE and concrete for which w/c was originally derived.

ACKNOWLEDGEMENTS

The authors would like to thank Robert Elliot and Simon Crispe for their efforts in conducting the experimental phase of this work. This project was supported by the UWA Faculty of Engineering, Computing and Mathematics.

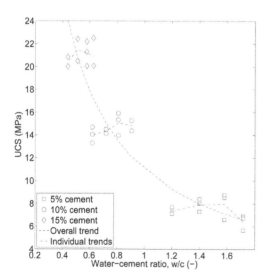

Figure 6. Specimen UCS vs water-cement ratio showing polynomial fit to data and individual material trends (from Figure 3).

REFERENCES

Beckett, C. & D. Ciancio (2014). Effect of compaction water content on the strength of cement-stabilised rammed earth materials. *Canadian Geotechnical Journal 51*(5), 583–590.

Bryan, A.J. (1988). Criteria for the suitability of soil for cement stabilization. *Building and Environment 23*(4), 309–319.

Bui, Q.-B., J.-C. Morel, S. Hans, & P. Walker (2014). Effect of moisture content on the mechanical characteristics of rammed earth. *Construction and Building Materials 54*, 163–169.

Ciancio, D., P. Jaquin, & P. Walker (2013). Advances on the assessment of soil suitability for rammed earth. *Construction and Building Materials* 42, 40–47.

Fernandes, V., P. Purnell, G. Still, & T. Thomas (2007). The effect of clay content in sands used for cementitious materials in developing countries. *Cement and Concrete Research* 37(5), 751–758.

Jaquin, P.A., C.E. Augarde, D. Gallipoli, & D.G. Toll (2009). The strength of unstabilised rammed earth materials. *Géotechnique* 59(5), 487–490.

Neville, A.M. (2011). *Properties of concrete* (5th Edition ed.). Pearson Education Ltd, Essex (UK).

Standards Australia (2003). AS1289.5.2.1.–2003. Methods of testing soils for engineering purposes. Method 5.2.1: Soil compaction and density tests Determination of the dry density/moisture content relation of a soil using modified compactive effort.

Venkatarama Reddy, B. & P. Prasanna Kumar (2010). Embodied energy in cement stabilised rammed earth walls. *Energy and Buildings* 42(3), 380–385.

Venkatarama Reddy, B. & P. Prasanna Kumar (2011a). Cement stabilised rammed earth. Part A: Compaction characteristics and physical properties of compacted cement stabilised soils. *Materials and Structures* 44, 681–693. 10.1617/s11527–010–9658–9.

Venkatarama Reddy, B. & P. Prasanna Kumar (2011b). Cement stabilised rammed earth. Part B: Compressive strength and stress-strain characteristics. *Materials and Structures* 44, 695–707. 10.1617/s11527–010–9659–8.

Rammed Earth Construction – Ciancio & Beckett (Eds)
© *2015 Taylor & Francis Group, London, ISBN 978-1-138-02770-1*

Contemporary soil-cement and rammed earth in South Africa

G. Bosman & K. Salzmann-McDonald
Earth Unit, Department of Architecture, University of the Free State, Bloemfontein, South Africa

ABSTRACT: Architects often explore the local vernacular built heritage for application in contemporary architecture. The past six decades some South African architects have designed and constructed several modernist buildings that reflect local materials in different climatic zones, even before the principals of critical regionalism was coined and popularized. These lessons learned from local traditional building can result in climatically well adapted, sustainable architecture that reflects the cultural context in a modernist language. In some cases superficial attempts using symbols and decorations of an ethnical cultural origin become decorative elements without a true representation of the materials that reflect the techniques used. These efforts illustrate the contemporary usefulness and advantages of available local material, local construction and developed skill of building trades. The aim of this paper is to illustrate the contribution of contemporary soil-cement after a short history of traditional rammed earth was introduced in South Africa. The construct process of these contemporary buildings support community development with-in the green cultural context. Through a critical analysis of these hybrid buildings, the authors argue that a critical regionalism approach do support community upliftment with in a modernist program by utilizing contemporary soil-cement construction.

1 INTRODUCTION

Lessons learned from past rammed earth building projects can result in improved, climatically well adapted, sustainable architecture that reflects a contemporary "Green" cultural context.

For many decades here has been the awareness in the built environment to learn from traditional and vernacular concepts for contemporary application (Foster, 1983). Furthermore, the new pillar of culture and governance is a welcome addition to the already considered social, economic and environmental pillars of sustainable development (Guillaud, 2010). Green building practises are not investigated enough, because people and organizations are used to routines and structures and are resistant to necessary change. The reasons are habitual routine, fear of the unknown and resource limitations (Hoffman & Henn, 2008).

Literature often reflects on vernacular architecture but seldom links this to the contemporary usefulness and advantages of the local material application to support sustainable architecture. Furthermore, the lack in knowledge, regarding energy conservation in relation to climate change, creates a lack of urgency in the need to develop sustainable green building practices. (Hoffman & Henn, 2008).

Earth construction techniques recorded for the second part of the 17th century in South Africa, used by ethnical groups and early colonial settlers (from the Dutch and British colonial epochs) utilized mainly natural stone, sods and adobe for the construction of wall elements. The introduction of rammed earth (traditionally un-stabilized *pisé de terre*) in southern Africa only arrived in the early 20th century from Europe. The authors would like to refer to soil-cement, since traditional European rammed earth was either un-stabilized or occasionally stabilized with 5–8% lime. Traditional rammed earth is not considered a local South African vernacular technique.

Soil-cement construction methods enjoy a growing presence in southern Africa in contemporary design and construction. Furthermore, the utilization of available local material can result in community upliftment and job creation that reflect social responsibility in the green cultural context. Lessons learned from newly appreciated earth construction materials and methods can support sustainable architecture through a holistic approach and application in contemporary buildings that surpasses the traditional architectural language of the context. This approach can provide a solution for a modernist program. Within the frame of critical regionalism (Foster, 1983) as explained by

Kenneth Frampton as a progressive approach that address the problems of the architecture of the International Style and post-modernism.

2 LEARNING FROM PAST EXPERIENCE

The utilization of earth as building material world wide (Fathy, 1973; Houben & Guilaud, 1994; Oliver, 2003; Seth, 1988) is well documented, as are local techniques in southern Africa (Fransen & Cook, 1965; Frescura, 1985). Even the European techniques adapted to suit local conditions in South Africa was documented during the 1920's when pise (rammed earth technique) was known in Zimbabwe after the introduction of this technique by French missionaries. This building technique spread throughout the Roman Empire and developed into a strong and well refined rammed earth culture in all the Mediterranean countries on three continents. Centuries later, early missionary expeditions brought this technique to southern Africa. South African literature can provide very little information of early experimentation with rammed earth construction techniques in this context.

Archibald et al. (1948) described the excellent office building examples in pisé built during 1920 at the City Deep Mine in Johannesburg. These buildings were constructed by a contractor from Zimbabwe. In 1922 an early farm house was constructed in rammed earth on the Daggafontein Farm, in the Springs district in Gauteng. During the same period the early station buildings were constructed in rammed earth at Simondium in the Western Cape Province. Other recordings of experimental rammed earth buildings in South Africa were noted in the Gauteng Province. Several buildings comprising of a small school, labourers' dormitories, mine offices, married and single quarters and other buildings were constructed at the Globe and Phoenix Mines in Gauteng during the 1940's (Archibald, et al., 1948).

2.1 Early SA research in rammed earth and soil-cement

In 1993 the thesis of Christian Roberg, entitled "The use of soil-cement as a construction material" contributed to the national research in earth construction when he chose three different soils from around Johannesburg and characterized their properties using soil science techniques. The soils he examined were quaternary sand with 14% clay and 3% gravel, as well as decomposed granite soil with 4% clay and 40% gravel in rammed stabilized earth walls. In both cases the compressive strength was above 5 MPa after 28 days of curing. Both soils required no more that 5% cement stabilization. (Morris & Booysen, 2000).

A critical analysis of selected South African contemporary soil-cement case studies provide more information as this popular technique supports the growing green building culture in southern Africa.

3 CASE STUDIES

3.1 Dalrymple pavilion: rech carstens architects

The Dalrymple pavilion project challenged the subcontractor (Insynch Sustainable Technologies) to achieve an experimental soil cement wall of exceptional quality on a difficult urban site.

The Dalrymple pavilion is built in the suburb of Westcliffe, Johannesburg. It is an entertainment pavilion and guest suite. On two levels the soil cement wall forms the stereo tomic screen of the building, forming the backdrop for the main lounge area, swimming pool and kitchen.

The wall drops down the steep site to the lower level, anchoring a smaller informal lounge and guest suite. The roof is supported by a steel and timber (railway sleepers) column and beam (techtonic) structure. The architects also made use of sealed mild steel plates at the entrance wall and gate. These natural materials are contrasted against the seamless white floors and kitchen cabinets. All these materials contribute to a very natural and organic yet sophisticated feel. Commercial quarried sands was transported to the site and stabilized with 8–10% cement.

Figure 1. The Dalrymple pavilion under construction (Photo: *Rech carstens architects*).

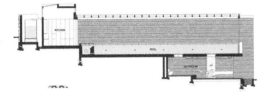

Figure 2. The Dalrymple pavilion longitudinal section (Photo: *Rech carstens architects*).

3.2 *Kathu office building: cube architects*

The Sishen Iron Ore Community (SIOC) Development Trust challenged the professional team of this office building in Kathu in the Northern Cape Province with a social upliftment and skills transfer agenda. The construction had to include a training program to support a local group of builders with the technical skills to start their own small business in soil-cement construction. The SIOC-CDT office building was generated from the landscape, climate and the social context of the Northern Cape. The semi-arid sandy savannah and the surplus of earth from mining activities as features from the natural and man-made environment, suggested soil cement as a viable solution (Oliver, 2013).

Insynch Sustainable Technologies were sub-contracted to train and supervise 18 members of the local Kathu community, who formed an upcoming business. The client supported this group with their own equipment to continue with other projects in the area. The design of the building aimed to capture the climatic and cultural context, by using the local skill, building technologies and materials. The language of mining was incorporated by the extensive use of steel in the design.

The engineers used 8–10% cement stabilisation for the different colour sands to improve the load bearing and surface erosion. The deep red coloured sand and gravel were from excavations on the site and other sands were sourced from nearby mine dumps and river.

After the completion of the office building the newly formed enterprise started on an extension of the existing buildings of the small airport of the local mine company. With this successful follow-up project the team is more established and technically proficient.

3.3 *Oliver Tambo education and narrative centre: Tunde Oluwa (principal) and eco design architects and consultants, Andy Horn*

This centre provided small business opportunities and in exchange educates the community about environmental issue (Momberg, 2012).

Figure 3. The roof and soil-cement walls of the SIOC-CDT office building. (Photo: Authors).

Figure 4. The care taker's cottage of the Oliver Tambo education and narrative centre. (Photo: Authors).

The OR Tambo Education and Narrative Centre is an environmental education centre which is an educational experiment that demonstrates green building practices, constructed on the banks of Leeupan, in Wattville-township, Gauteng. Many different alternative technologies were used, such as a trombe wall (a sun facing wall which absorbs heat during the day and slowly releases it into the room at night), earth floors, wind catchers and straw bale walls plastered with cow dung. Stone and urbanite walls (recycled from industries in the area) were also used. Roof gardens were used to assist with cooling. Rain water is harvested, grey water is recycled and solar energy is used for heating via earth tubes (Momberg, 2012).

4 CONCLUSIONS

The selected case studies in different parts of South Africa together with other green building projects all work together to sensitize the South African public. Not only does it promote using local materials, cutting transportation costs to the minimum, it also encourages local skills development to enrich knowledge and create new sustainable skills. These "environmental concerns are integrated into the existing routines by which buildings are constructed, recasting them in ways that are mutually beneficial to the objectives of individuals, organizations and the sustainability of the ecosystems on which they depend" (Hoffman & Henn, 2008).

The Dalrymple pavilion did not aim to facilitate community development, but this project gave the soil cement contractor the opportunity to experiment and refine the construction technique. This preceded the SIOC CDT office building project in which the same contractor had to accommodate skill and knowledge transfer. Therefore the Dalrymple pavilion acted as a proxy to community development in regard to soil-cement as technology. The OR Tambo Education and Narrative Centre project was on hold for two years until 2010. During this period the professional team picked up on experience and knowledge regarding "green building" and sustainable construction. These contemporary buildings are worthy examples of soil-cement in South Africa.

REFERENCES

Archibald, A.J., Crosby, A.J.T., & Patty, R.L. (1948). Cheap building by *pisé de terre* methods. SA Institute of Race Relations. *The Natal Witness*, Ltd, Pietermaritzburg.

Fathy, H.C. (1973). *Architecture for the poor: An experiment in rural Egypt*. Chicago: University of Chicago Press.

Foster, H. (1983). The Anti-Aesthetic: Essays on Postmodern Culture, Seattle: Bay Press.

Fransen, H. & Cook, MA. (1965). *The old houses of the Cape: A survey of the existing buildings in the traditional style of architecture of the Dutch-settled regions of the Cape of Good Hope*. Cape Town: Balkema.

Frescura, F. (1985). Rural shelter in southern Africa. Johannesburg: University of the Witwatersrand.

Guillaud, H. (2010). Proceedings of the TERRA Education-seminar—workshop, CRAterre-ENSAG: Grenoble.

Hoffman, A.J. & Henn, R. (2008). Overcoming the Social and Psychological Barriers to Green Building. *Organization and Environment*, Volume 21. Number 4, Dec. 2008. 390–419.

Houben, H. & Guillaud, H. (1994). *Earth construction—A comprehensive guide*. London: Intermediate Technology Publications.

Momberg, E. (2012). Tribute to a hero. *Earthworks*. Issue 7, April—May. 53–60.

Morris, J. & Booysen, Q. (2000). Earth Construction in Africa. Paper presented at the Proceedings of the *Strategies for a Sustainable Built Environment Conference*, Pretoria (23–25 Aug.)

Oliver, P. (2003). *Dwellings: The Vernacular Houses World Wide*. London: Phaidon.

Oliver, P. (ed). (1969). *Shelter and society*. London: Barrie & Rockliff: The Cresset Press.

Olivier, J. (2013). Office building, Kathu—Casting a tree-like silhouette in the Kalahari. *Journal of the South African Institute of Architects*, Issue 62, July/Aug 2013, 20–23.

Seth, S. (1988). *Adobe! Homes and interiors of Taos, Santa Fe and the Southwest*. Stamford, Connecticut: Architectural Book Pub. Co.

Rammed Earth Construction – Ciancio & Beckett (Eds)
© 2015 Taylor & Francis Group, London, ISBN 978-1-138-02770-1

The creep of Rammed Earth material

Q.B. Bui
LOCIE, CNRS, Polytech Annecy-Chambéry, University of Savoie, Savoie, France

J.-C. Morel
LGCB, LTDS, CNRS, ENTPE, University of Lyon, Lyon, France

ABSTRACT: In this study, the creep of Rammed Earth (RE) material are studied on the walls which have been constructed and exposed for 22 years to natural weathering. First, mechanical characteristics of the "old" walls were determined by two approaches: in-situ dynamic measurements on the walls; laboratory tests on specimens which had been cut from the walls. Then, the walls' soil was recycled and reused for manufacturing of new specimens which represented the initial state. Comparison between the mechanical characteristics of the walls after 22 years on site and that of the initial state enables to assess the walls' creep.

1 INTRODUCTION

Rammed Earth is attracting renewed interest thanks to its "green" characteristics. Several research investigations have recently been conducted to study the characteristics of RE, however, to our knowledge, there is not yet any scientific study on its aging. Indeed, the famous phenomenon relative to the aging of old RE structures is the buckling.

Figure 1 presents an example of a RE wall in France which is more than 200 years old. Note that the shadow of the roof on the front wall is not a horizontal straight line. This is due to the horizontal buckling of this wall that can be seen on the plan. The buckling phenomenon depends on several parameters (wall's slenderness ratio, eccentricity of the loading, boundary conditions...) but an important parameter which relates to material's characteristic is the creep phenomenon.

Shadow of the roof on the front wall

Figure 1. A Rammed Earth wall more than 200 years old.

2 CHARACTERIZING 22 YEARS OLD WALLS

2.1 Presentation of studied walls

The walls were built in 1985 thanks to the Rexcoop program, controlled by the French Scientific and Technical Building Center (CSTB), near Grenoble, in a French Alpine valley, at an altitude of 212 m. The temperature of the site can vary from −20°C to 38°C for some particular years and its average varies from 2°C to 20°C. The annual rainfall is about 1000 mm, the direction of prevailing winds is NE-SW and the maximum wind speed is 21 m/s.

The walls (1 m width × 1.1 m height × 0.4 m thickness) were manufactured on a concrete foundation with a 25 cm base exposed above ground level. A bituminous layer was painted on top of the base to prevent the capillary rise. The walls were protected by asbestos cement roof. More information can be found in Bui et al. 2009a.

The manufacturing water content of the soil was about 10%. There was no control of the walls' density.

2.2 In-situ dynamic measurement

2.2.1 Measurement device
Three accelerometers with a sensitivity of 1 μg (with g being the gravity field) were placed on top of the wall: two accelerometers in the centre to measure two horizontal accelerations following the two main axes of the wall; and another accelerometer on the edge to measure possible torsional movements (Figure 2).

Figure 2. Arrangement of the sensors on the wall A.

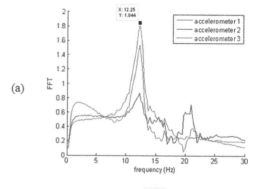

(a)

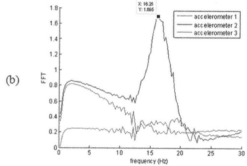

(b)

Figure 3. A result of wall A induced by shock 1(a), shock 2(b).

The excitation consisted in a light shock (by a hammer) which was applied to the top of the wall. Three configurations were carried out: (1) a central shock following the transversal direction of the wall, (2) a central shock following the longitudinal direction of the wall, (3) an offset shock following the transversal direction. These configurations excite the possible vibration modes of the wall: transversal, longitudinal and torsional.

2.2.2 Frequencies measured in-situ

Three RE walls were measured, 22 years after their construction. They had three different types of protection: the reference RE—with no protection layer (wall A), the RE protected with plaster (wall B) and the RE protected with paint (wall C). Since the wall B was protected with plaster (thickness about 3–4 cm), the measurements were made after removing the plaster to eliminate its influence on the results. For the wall C protected with paint (thickness < 0.5 mm), the measurements were performed without removing the protection, assuming that its contribution to the wall's dynamic behaviour was negligible.

Figure 3 shows typical results obtained after a signal processing for the shocks 1 and 2. Each peak corresponds to a modal frequency. In this case the frequency of the first transversal mode is identified at 12.25 Hz. For shocks 1, sensor 2 does not give a clear signal because in this case, excitation was perpendicular to the sensor 2; there was no major vibration in the wall's longitudinal direction. The result of sensor 3 is similar to that of sensor 1, since the torsion was not clearly captured (shocks were not important enough to solicit this mode). For shock 2, sensors 1 and 3 did not give significant information but sensor 2 captured the second vibration mode which was in the longitudinal direction. In this case the second modal frequency is of 16.25 Hz. Results of two others walls are presented in Table 1.

2.2.3 Finite Element Modelling (FEM) of the walls

The walls are modelled with solid elements. For given dimensions, the wall's natural frequencies

Table 1. Frequencies measured and corresponding moduli identified by the model.

Wall	E (MPa)	Measurements		Model	
		f_1 (Hz)	f_2 (Hz)	f_1 (Hz)	f_2 (Hz)
A	104	12.25 ± 0.05	16.25 ± 0.12	12.25	16.48
B	98	11.87 ± 0.08	16.20 ± 0.09	11.87	15.97
C	90	11.38 ± 0.10	15.38 ± 0.22	11.38	15.31

depend only on its density and its elastic characteristics. For the modelling, the material was assumed isotropic (which is acceptable for dynamic measurements which were performed in very small strain, Bui et al. 2009b). The Poisson's ratio was taken of 0.22 following Bui et al. study (2014). The link between the RE and the concrete foundation is considered to be an embedment. The principle to determine modulus from natural frequencies was presented in Bui et al. 2009b.

The results of the identified modulus are given in Table 1. From the results presented in Table 1, the Young's moduli of the measured walls were about of 100 MPa. These values are lower than that indicated in the literature (for example, Bui et al. 2009b had identified moduli about 450 MPa). This value of the

Young's modulus will be compared in the next section with that obtained by compression tests.

3 LABORATORY STATIC TESTS

3.1 Cutting out the specimens

The specimens were taken from the walls using a chainsaw generally used to cut concrete and stones. The disadvantages of using this method to take specimens are the generation of vibrations and the use of water, which can decrease the mechanical strength of the specimens, notably due to the splitting apart of the layers of earth.

3.2 Density measurement

Since the specimens taken from the walls were roughly shaped, their density was estimated using hydrostatic weighing, once they had been coated with paraffin to seal them. Three specimens approximately 8 cm × 10 cm × 20 cm in size were measured which gave a mean dry density of 1.82 ± 0.01. The dry density value of 1.82 is lower than that of the new RE presented in a previous study, Bui et al. 2009, which was around 1.92. This lower dry density will be discussed in the next sections.

3.3 Unconfined compression test

The specimens cut from the walls were transported to the laboratory and re-shaped with a table saw. Several specimens were cut from the walls, but only six specimens had an acceptable quality for the testing. Each specimen had dimensions of 16 cm × 16 cm × 27 cm (slenderness ratio of 1.7). Three were tested in the direction perpendicular to the layers, and three others in the direction parallel to the layers.

Table 2 gives the results of this test. The modulus is calculated for stress levels between 0 and 20% of the maximum stress which represent the elastic part of RE material (Bui et al. 2014). There is no important difference in moduli between the vertical and the horizontal directions of the wall (E ≈ 95 MPa). The strength of the specimens tested in the direction parallel to layers is slightly less than that of the specimens tested in the vertical

Table 2. Results obtained from the compression tests.

Direction (compared to layers)	Moisture content	Stress max (MPa)	E (MPa)
Perpendicular	1.4%	0.89 ± 0.10	98 ± 6
Parallel	1.3%	0.82 ± 0.08	93 ± 5

direction (8%). No major difference in two directions' moduli shows that the isotropic hypothesis which was assumed in the FEM (for small strains) is acceptable. This remark was noted in previous studies (Bui et al. 2009b, c, 2014).

A good correlation of the moduli obtained by the dynamic and static methods can be observed, which are around 100 MPa, that confirms the relevance of the results obtained.

4 CHARACTERIZING THE "NEW" RE

4.1 Manufacture of specimens and compression tests

The soil of the specimen cut from the walls was recycled and reused for the manufacturing of the "new" specimens. In order to test the specimens in two directions (perpendicular and parallel to the layers), two types of specimens were manufactured:

- Two specimens 0.4 m × 0.4 m × 0.7 m
- Two specimens (0.4 × 0.4 × 0.2) m³

Discussions about the representativeness of specimens manufactured in laboratory were presented in the literature (Bui et al. 2009b,c). Indeed, to ensure a faithful representation of the in-situ wall material, the manufacturing mode and material used for laboratory specimens should be as identical as possible to those used in situ. The manufacturing water content and the compaction energy in the laboratory were chosen similar to that on site. The manufacturing water content was 10%. The dimensions of specimens tested in the direction perpendicular to the layers were 40 cm × 40 cm × 70 cm, with nine layers.

The specimens to be tested in the parallel direction were composed of three layers; their dimensions were 40 cm × 40 cm and 20 cm high. Special attention was given during compaction of the last layer to obtain a surface that was as flat as possible. To achieve a slenderness ratio of 2, the specimens were then cut with a table saw. Two specimens 40 cm × 40 cm × 20 cm provided four specimens 20 cm × 20 cm × 40 cm for testing in the parallel direction. For specimens tested in the direction parallel to the layers, surfacing was not necessary, because the two faces that were in contact with the formwork were sufficiently flat.

4.2 Results

The results are presented in Table 3. Once again, there is no important difference between the elastic moduli of the vertical and horizontal directions of the wall (E ≈ 270 MPa). The compressive strength

Table 3. Compression tests on the new specimens.

Direction (compared to layers)	Moisture content	Stress max (MPa)	E (MPa)
Perpendicular	1.8%	1.35 ± 0.1	263 ± 12
Parallel	1.7%	1.18 ± 0.1	287 ± 8

of the specimens tested in the direction parallel to layers is slightly less than that of the specimens tested in the perpendicular direction (8%).

5 DISCUSSION AND THE CREEP COEFFICIENT

Except for the case of exterior walls, the in-situ walls in this study are not directly representative of current RE houses. Indeed, they have two outer surfaces, whereas in a house, the interior atmosphere (without rain) is different from the external one subjected to weathering.

The strategy which studies the initial state by manufacturing the new specimens from the recycled soil is questionable because the new ones are similar but not the same as the initial state ones. On one hand, are there possible changes in the soil's characteristics after 22 years on site due to cycles of adsorption-desorption, freeze-thaw? On the other hand, the new specimens have a higher dry density than that of the old walls. There are two possible reasons for this: firstly, compaction energy which depends on the artisan experience is higher in the case of the new specimens. Secondly, there is a possible change in the porosity of the walls due to the adsorption-desorption, freeze-thaw cycles (Grossein 2009). If the second reason is confirmed, it will be an interesting element in investigations on the ageing effects of rammed earth walls.

Due to the difference of the dry density (about 5%) between the old walls and the new specimens, there is a difference in the corresponding compressive strength (about 50%). This result is not surprising and was observed in the literature (Morel et al. 2007). The case of the Young's moduli is different. In the present study, the new specimens had moduli which were 2.7 times greater than the ones of the old walls. Besides the above reasons, another possible reason is the creep phenomenon. The creep is the tendency of a material to deform permanently under the influence of loadings (although the stress does not increase). For concrete, the creep occurs at all stress levels and, within the service stress range, is linearly dependent on the stress if the pore water content is constant (Eurocode 2). For RE, the increase of strain under a constant

stress was noted in Lombillo et al. study (2014) and compared to the phenomenon of consolidation of normally consolidated soil.

The creep depends on the ambient humidity, composition of the RE material, age of material, duration and intensity of the loading. For concrete, following Eurocode 2, the effective modulus E_{eff} is related to the initial modulus E_{t0} (at 28 days) by the formula:

$$E_{eff} = E_{t0} / [1 + \varphi] \qquad (1)$$

where φ is the creep coefficient:

$$\varphi = E_{t0} / E_{eff} - 1 \qquad (2)$$

If the modulus of the walls and the new specimens are respectively used for E_{eff} and E_{t0}, the corresponding creep coefficient of the walls is 1.7. This information is interesting because to our knowledge, this is the first time a value of RE creep coefficient is presented.

In the last decade, to explain the basic creep of concrete, physical mechanisms taking action at the scale of the hydrates were proposed; they are based on the microprestress-solidification theory, the viscoplastic behaviour of the hydrates (principally the C–S–H which is the principal component of the cement), and the rearrangement of nanoscale particles (C–S–H level) following the free-volume dynamics theory of granular physics (a synthesis can be found in Rossi et al. 2014). However, in the case of RE, these theories cannot explain its creep because there are not C–S–H particles. Rossi et al. (2014) showed that even others physical mechanisms can exist, the main physical origins of the basic creep are related to the microcracking propagation under load. In the case of RE, when a wall is under a loading (self-weight, wind, temperature, freeze-thaw), it cracks. The microcrack propagation is suggested to be the main factor for the decrease of the Young's modulus of RE material.

6 CONCLUSIONS AND OUTLOOK

The strategy which studies the initial state by manufacturing the new specimens from the recycled soil is questionable. However, it is always interesting to have direct information from real walls exposed to natural conditions. Following the obtained results, the ageing has a significant effect on the Young's modulus but not on the compressive strength. So, for buckling studies on old RE walls, the creep coefficient should be taken into account. This is a first exploratory study on the ageing and creep of RE walls, so the results should be confirmed by other studies in the future.

REFERENCES

Bui Q.B., Morel J.C., Reddy B.V.V., Ghayad W. 2009a. Durability of rammed earth walls exposed for 20 years to natural weathering", *Building and Environment,* 44: 912–919.

Bui Q.B., Morel J.C., Hans S., Meunier N. 2009b. Compression behaviour of nonindustrial materials in civil engineering by three scale experiments: the case of rammed earth, *Materials and Structures,* 42: 1101–1116.

Bui Q.B., Morel J.C. 2009c. Assessing the anisotropy of rammed earth, *Construction and Building Materials,* 23: 3005–3011.

Bui Q.B., Morel J.C., Hans S., Walker P. 2014. Effect of moisture content on the mechanical characteristics of rammed earth, *Construction and Building Materials,* 54: 163–169

Eurocode 2, 2005. Design of concrete structures—part 1–1: general rules and rules for buildings, *NF EN 1992–1-1, P 18–711–1.*

Grossein O. Modélisation et simulation numérique des transferts couplés d'eau, de chaleur et de solutés dans le patrimoine architectural en terre, en relation avec sa dégradation, *PhD thesis (in French)*, Université Grenoble 1, 2009, 226p.

Lombillo I., Villegas L., Fodde E., Thomas C. 2014. In situ mechanical investigation of rammed earth: Calibration of minor destructive testing, *Construction and Building Materials*, 51: 451–460.

Morel J.C., Pkla A., Walker P. 2007. Compressive strength testing of compressed earth blocks, *Constr Build Mater*, 21: 303–309.

Rossi P., Charron J.P., Bastien-Masse M., Tailhan J.L., Le Maou F., Ramanich S. 2014. Tensile basic creep versus compressive basic creep at early ages: comparison between normal strength concrete and a very high strength fibre reinforced concrete, *Materials and Structures* 47:1773–1785.

Rammed Earth Construction – Ciancio & Beckett (Eds)
© 2015 Taylor & Francis Group, London, ISBN 978-1-138-02770-1

Discrete element modeling of Rammed Earth walls

Q.B. Bui
LOCIE, CNRS, University of Savoie, Savoie, France

T.-T. Bui & A. Limam
LGCIE, INSA Lyon, Lyon, France

J.-C. Morel
LGCB, CNRS, ENTPE, France

ABSTRACT: Several research studies have recently been carried out to investigate Rammed Earth (RE) material. Some of them attempted to simulate the RE's mechanical behavior by using analytical or numerical models. These studies always assumed that there was perfect cohesion at the interface between earthen layers. This hypothesis proved to be acceptable for the case of vertical loading, but it could be questionable for horizontal loading. To address this problem, Discrete Element Modeling (DEM) seems to be relevant to simulate a RE wall. To our knowledge, no research has been conducted thus far using DEM to study a RE wall. This paper presents an assessment of the DEM's robustness for RE walls. Thirteen parameters that were necessary for DEM were identified. The relevance of the model and the material parameters were assessed by comparing them with experimental results from the literature.

1 INTRODUCTION

Several investigations have recently been conducted to study the mechanical characteristics of RE by using analytical or numerical methods (Bui et al. 2009, 2011, 2014b, Ciancio & Augarde 2013, Ciancio & Robinson 2011). These studies assumed that the cohesion at the interface between earthen layers was perfect. This hypothesis proved to be acceptable for the case of vertical loading (Bui et al. 2014b), but it could be questionable for horizontal loading. To address this problem, Discrete Element Modeling (DEM) seemed to be a relevant means of simulating a RE wall. To our knowledge, no study has used DEM to study RE walls. Therefore, this paper presents an assessment of the DEM's robustness in the case of RE walls.

2 DISCRETE ELEMENT METHOD

The 3DEC code (Itasca 2011) is used in this study. The RE wall was modeled as an assemblage of discrete blocks (earthen layers), and the interfaces between earthen layers were modeled by introducing an interface law.

Earthen layers were assumed to be homogeneous, isotropic and were modeled by blocks which were further divided into a finite number of internal elements for stress, strain, and displacement calculations. The failure envelope used in this study was the Mohr-Coulomb criterion with a tension cut-off. Interfaces between earthen layers were modeled by an interface law between the blocks following the Mohr-Coulomb interface model with a tension cut-off. This interface constitutive model considers both shear and tensile failure, and interface dilation is included. More details of the constitutive behaviors of blocks and interfaces can be found in Bui et al. (2014c).

3 IDENTIFICATION OF PARAMETERS

Thirteen parameters need to be identified for the model:

– Earthen layer (7 parameters): density, Young modulus, Poisson's ratio, tensile strength, cohesion, friction angle, and dilatancy angle;
– Interface (6 parameters): normal stiffness, tangent stiffness, tensile strength, cohesion cinterface, friction angle, and dilatancy angle.

Among the aforementioned parameters, several characteristics of the earthen layer can be directly or indirectly found in the literature (density and Young modulus: Bui et al. 2009, 2014a; Poisson's ratio: Bui et al. 2014a; tensile strength and cohesion: Bui et al. 2014b; friction angle: Cheah et al. 2012; Bui et al. 2014b; and dilatancy angle:

Vermeer et al. 1984), which will be detailed in the next sections. However, parameters concerning the interface have not been reported previously.

3.1 Identification of the interface's elastic stiffness

Results from in situ non-destructive tests in Bui et al. (2009) were used to identify the interface's elastic stiffness. In that study, the wall's natural frequencies were identified which enabled to determine the wall's Young modulus.

The wall was modeled by DEM; its density, Young modulus and compressive strength were of 19.5 kN/m³, 470 MPa and 1.0 MPa, respectively. The Poisson's ratio of 0.22 reported by Bui et al. 2014 was applied. Other parameters are less important in this case because the strains were small during dynamic measurement; their identification will be presented in detail in the following section of this paper. A parametric study was performed. The value of k_n (so k_s), which reproduced the first measured frequency (the most important frequency), was selected; the correlation of other frequencies was also checked (more details can be found in Bui et al. 2014c). A stiffness of $k_n = 60$GPa/m (so $k_s = 24.6$GPa/m) was selected.

3.2 Identification of the earthen layer's parameters

The walls in a previous study (Bui et al. 2014b) were used to identify the parameters in compression. These walls measuring $(100 \times 100 \times 30)$ cm³ were subjected to concentrated loads on a (30×30) cm² surface at the middle of the wall parameters.

In order to assess the influence of interfaces, two DEM models were studied: without interface (homogeneous material) and with interfaces. The contact between the RE wall and the concrete base was modeled by a contact which follows the Mohr-Coulomb law and had a friction angle of 15° (usual value for concrete surface).

3.3 Influence of interfaces in compression test

Experimental compressive and tensile strengths of earthen layers were used (f_c and $f_{t,layer}$ respectively). For other parameters, a parametric study was performed. The synthesis of the parameters obtained is presented in Table 1.

In order to assess the influence of the interface in this case, low values of cohesion, tensile strength, and friction angle of interfaces were chosen for the model, which were 25% of the respective values of earthen layers. The cohesion and friction angle of earthen layers were chosen of $0.1f_c$ and 45°, respectively. The sensitivity of these parameters

Table 1. Parameters of earthen layer and interface used in the case of vertical loading.

	Tensile strength	Cohesion	Friction angle	Dilatancy angle
Layer	10%f_c	10%f_c	45°	12°
Interface	25% $f_{t, layer}$	25% c_{layer}	25°	12°

Figure 1. Failure in experiment (left, F_{max} = 118 kN), model without (middle, F_{max} = 120 kN) and with interfaces (right, F_{max} = 120 kN).

will be discussed in the next section. The results are presented in Figure 1. These results show that even with very low interface parameters, the results obtained by models with or without interfaces were similar which means that the role of interfaces does not important in the case of vertical loading.

3.3.1 Assessing the cohesion and friction angle of earthen layers

As the interface could be neglected in this case, only the influence of the cohesion and friction angle of earthen layers was studied. Two cases were considered to optimize the numerical model:

– Friction angle was fixed at 45°; cohesion varied from 7% to 10% of the compressive strength. The results are presented in Figure 2.
– Cohesion was fixed at 7% and 9% of the compressive strength, respectively; friction angle varied from 45 to 56°. The results are presented in Figure 3.

From Figures 2 and 3, the best cohesion value for the layers is in the range 7 to 9%. Figure 3 shows that the friction angle does not play a significant role for the first crack load (which corresponds to the slope's change in the load-displacement curve). However, it does affect the ultimate load. Figure 2 shows that the cohesion has an effect on the first crack load and the ultimate load. When this parameter increases, the first crack and ultimate loads increase; the more important effect on the ultimate load can be observed.

From these results, the best pair of layer characteristics was $c_{layer} = 9\%$ f_c and $\varphi_{layer} = 46°$, which gave the results closest to the experimental results: first crack load and ultimate load. There were still some differences to the experimental results, which may

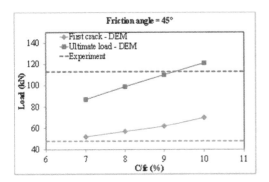

Figure 2. Influence of layer cohesion on the first crack load and ultimate load.

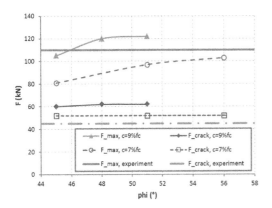

Figure 3. Variation of first crack load and ultimate load in function of friction angle.

come from the limit of the models used for DEM (Mohr-Coulomb for earthen layers, tension cut-off for interfaces). However, the study could confirm the results presented in previous studies in which the layer's cohesion was about 7–10% of compressive strength and the layer's friction angle was about 46–51° (Bui et al. 2014b, Cheah et al. 2012).

3.4 Identification of cohesion and tensile strength of the interface

Results of diagonal compression tests presented in Silva et al. (2013) were used. In that study, walls of $55 \times 55 \times 22$ cm³ compacted in 9 layers were tested. With a diagonal compression, the interface shear behavior was solicited more than for vertical loading, which made it possible to determine the interface characteristics (Figure 4).

The wall GSRE_7.5 was studied, i.e., granitic stabilized rammed earth with 7.5% of ash (by weight). In order to assess the influence of interfaces, several cases were studied:

Figure 4. Model with and without interfaces.

– Cohesion, tensile strength, and friction angle of interfaces were the same as earthen layers. This is equivalent to the case without interface.
– Cohesion, tensile strength, and friction angle of interfaces of 85% and 90% of the earthen layers, respectively, were taken. A preliminary study showed that lower values for interface parameters could not reproduce the experimental results in this case.

3.4.1 Model without interface

Following the above results, the layer cohesion was 7–10% of the compressive strength; the layer friction angle varied from 45–56°. From Figure 5, all models could reproduce the shear modulus of the wall, which was the slope of the first part of the test (until 20% of the ultimate shear stress). However, these models could not reproduce the slope's change after the first part of the test. Among these results, cohesion $c_{layer} = 10\%f_c$ and friction angle $\varphi_{layer} = 50°$ gave the most adapted result for the ultimate shear strength. This model could reproduce the diagonal crack of the experiment (Figure 6) but it could not yet reproduce the horizontal crack at the interface between the third and fourth layers (from the top).

3.4.2 Model with interfaces

Several values were tested for the interfaces: interface cohesion and interface friction angle were respectively of 70, 80, 85 and 90% of the corresponding parameters of the earthen layer.

The interface elastic stiffness which was determined in a previous part of the study was used. The results are presented in Figure 7. It is observed that the models with interfaces could reproduce the first part of the experimental result up to 50% of the ultimate shear stress, which was better than the model without interface. Among the models with interfaces, the model in which the interface's characteristics constituted 90% of the layers gave results closest to those of the experimental model. Indeed, the peak for this model was closest to that of the experimental curve, both for shear stress and shear distortion.

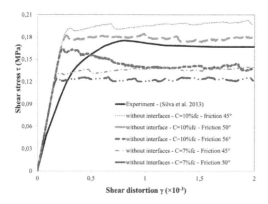

Figure 5. Results of models without interfaces and those of the experiment.

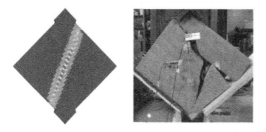

Figure 6. Model without interface and experiment.

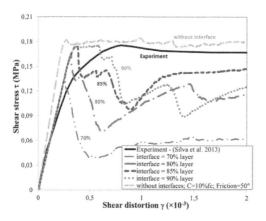

Figure 7. Comparison of numerical and experimental results.

However, none of the studied models could reproduce the nonlinear behavior of the experiment before the peak. This can be explained by the behavior law used for the interfaces, which is linear before the failure. Nevertheless, when the failure modes were observed (Figure 8), the model with interfaces at 85% of the layer's characteristics was the one that could reproduce the failure at the interface of the third layer. The model with interfaces

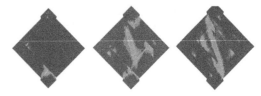

Figure 8. Models with interfaces $c_{interface}$ = 70 (left); 85 (middle) and 90% c_{layer}.

Figure 9. Experimental failure mode of GSRE_2.5 and 5.0.

having 90% of the layer's characteristics could only reproduce the diagonal crack. For the model with interfaces having 70% of the layer's characteristics, the failure was a local failure of the interface that was not representative of the experiment. These results show that a discrete model with interfaces having 85–90% of the layer's characteristics is the most relevant in this case.

3.5 Discussion of the interface characteristics

The results of the models in which the interface's characteristics represented 90–100% of the layer's characteristics could reproduce the slope of the elastic part, the ultimate shear stress, and the diagonal crack of the experiment. This failure mode was observed on other walls in the study of Silva et al. (2013), Figure 9. This means that the interface's characteristics can vary from 85% to 100% and still give satisfactory results.

4 CONCLUSIONS AND PROSPECTS

In this paper, identifying thirteen parameters of the DEM was presented. The following values can be suggested for RE walls:

– For earthen layers, considered homogeneous and isotropic: the tensile strength and the cohesion can be taken of about 10% of the compressive strength of the earthen layer. This result is similar to that presented in Bui et al. 2014b. The friction angle of about 50° can be taken. The dilatancy angle can be taken at 12° but this parameter did not play an important role in the case under in-plane loading. However, it would be interesting to study the importance of this parameter for out-plane loading (wind, earthquake).

- For interfaces: cohesion, tensile strength, and friction angle should be taken of about 85–100% of the respective values of earthen layer. A dilatancy angle of 12° can be taken. Interface stiffness of $k_n = 60$ GPa/m can be taken.

Although the model used in this paper could not reproduce the nonlinear behavior before the peak, it could reproduce others aspects of the different tests (ultimate load, failure mode).

REFERENCES

Bui Q.B., Morel J.C., Hans S., Meunier N. 2009. Compression behaviour of nonindustrial materials in civil engineering by three scale experiments: the case of rammed earth, *Materials and Structures*, 42, 1101–1116.

Bui Q.B., Morel J.C., Hans S., Walker P. 2014a. Effect of moisture content on the mechanical characteristics of rammed earth, *Construction and Building Materials*.

Bui T.T., Bui Q.B., Limam A., Maximilien S. 2014b. Failure of rammed earth walls: from observations to quantifications. *Construction and Building Materials*, 51, 295–302.

Bui T.T., Bui Q.B., Limam A., Morel J.C. 2014c. Modelling rammed earth wall by discrete element method, *Engineering Structure*, (in revision).

Cheah J.S.J., Walker P., Heath A., Morgan T.K.K.B. 2012. Evaluating shear test methods for stabilised rammed earth, *Construction Materials*, Vol. 165, Issue CM6, 325–334.

Ciancio D., Robinson S. 2011. Use of the Strut-and-Tie Model in the Analysis of Reinforced Cement-Stabilized Rammed Earth Lintels, *Journal of Materials in Civil Engineering*, Vol. 23, No. 5.

Ciancio D., Augarde C. 2013. Capacity of unreinforced rammed earth walls subject to lateral wind force: elastic analysis versus ultimate strength analysis, *Materials and Structures*, Vol. 46, 1569–1585.

Itasca, 3DEC—Three Dimensional Distinct Element Code, Version 4.1, *Itasca*, Minneapolis, 2011.

Silva R.A., Oliveira D.V., Miranda T., Cristelo N., Escobar M.C., Soares E. 2013 Rammed earth construction with granitic residual soils: The case study of northern Portugal, *Construction and Building Materials*, Vol. 47, 181–191.

Vermeer P. A., de Borst R. 1984. Non-associated plasticity for soils, concrete and rock. *Heron* 29, 1–64.

Rammed Earth Construction – Ciancio & Beckett (Eds)
© *2015 Taylor & Francis Group, London, ISBN 978-1-138-02770-1*

Recycled materials to stabilise rammed earth: Insights and framework

J.A.H. Carraro
Centre for Offshore Foundation Systems, The University of Western Australia, Perth, Australia

ABSTRACT: From a fundamental soil mechanics standpoint, rammed earth is an unsaturated compacted soil mixture that can have inter-particle bonding and reinforcing elements. The mechanical behavior of rammed earth can thus be interpreted using a rigorous geomechanics framework. In this paper, fundamental concepts required to analyse stabilised soil mixtures are outlined, with a particular focus on soil fabric, cementation and fiber-reinforcement issues. Assessment of specific gravity, particle shape, particle size distribution, plasticity, and volumetric indices of all fractions in a soil mixture can be useful for the rational (non-empirical) design of such materials, including rammed earth. Recent research on the beneficial use of waste materials in soil stabilisation is summarised with a focus on techniques aiming at increasing stiffness and strength, mitigating swell potential and improving thermal conductivity of stabilized soil mixtures. The successful use of waste materials such as carbide lime, scrap tire rubber, waste fibers, and various types of fly ash in soil stabilisation applications is highlighted. The potential use of alternative waste materials within a rational and mechanistic design framework that may be useful for rammed earth is discussed.

1 INTRODUCTION

From a geotechnical engineering perspective relying on fundamental soil mechanics, rammed earth is an unsaturated, compacted mixture of soils that can present inter-particle bonding and/or fiber reinforcement. Conceptually, the mechanical behavior of such materials can be interpreted using a rigorous geomechanics framework. In practice, modeling of geomaterials featuring (1) transitional soil behavior, (2) cementation, (3) fiber reinforcement and/or (4) partial saturation is complex. If considered separately, each of these features makes analyses of such geomaterials very challenging, never mind dealing with all of them at the same time.

Transitional soil behaviour refers to the response of 'intermediate' soils or soil mixtures whose mechanical response lies between the expected responses of classical soils such as clean sands and pure clays (Martins et al. 2001). Such soils are largely influenced by additional factors such as soil fabric and partial drainage, in addition to classical state variables such as density and stress.

Cementation, a term broadly used to refer to inter-particle bonding, affects soil fabric stability, thus increasing the realm of possibilities for arrangements among soil particles of various sizes, shapes and mineralogy (Mitchell and Soga 2005).

The presence of discrete inclusions in soils such as fibers or fibrous assemblies places the resulting mixture within the context of composite materials. The mechanical response of composite materials is complex and dependent on the amount, distribution and characteristics of all individual fractions present in the composite (Hull and Clyne 1996).

The proper approach to support mechanistic analyses of partially saturated soil can only derive from the adoption of rigorous unsaturated soil mechanics principles (Fredlund and Rahardjo 1993). Unsaturated soil mechanics has made great progress in recent decades, although its implementation into geotechnical practice may still lag behind possibly due to the higher level of complexity associated with the assessment of partial saturation. Such a comprehensive treatment is out of the scope of this paper so features 1–3 will be discussed next by assuming soil (or soil mixtures) of interest are saturated.

Stabilised rammed earth features most (or all) of the above issues, placing this material among the most complex geomaterials found in nature. While this allows for an honest appreciation of the true complexities associated with analyses of the mechanical behavior of rammed earth, stabilised rammed earth is a man-made material. As such, this should allow for a high degree of control and make engineering design of such material possible as long as fundamental mechanisms affecting its mechanical behaviour are well understood. A more recent challenge facing 21st century engineers is to understand and manipulate the mechanics of materials from an environmentally friendly perspective. This constitutes the main goal of this paper: to assess the mechanical behavior of soil mixtures including waste materials from a sustainable and yet mechanistically sound standpoint. Insights into rammed earth design are also highlighted.

2 CONCEPTUAL FRAMEWORK

2.1 *Behaviour of saturated, disturbed soil*

The mechanical behaviour of saturated, disturbed soil (i.e., soil with void space completely filled by water and showing fabric-independent response) can be predicted based on the unique relationship existing between density and stress state variables at critical state (Schofield and Wroth 1968). If a few soil parameters are known, knowledge of one variable allows the other to be assessed. For saturated, disturbed soil, this framework has been successfully used to model, for example, soil behaviour toward critical state through

$$q\delta\varepsilon_q^p + p'\delta\varepsilon_p^p = Mp'\delta\varepsilon_q^p + f_1 + f_2 + f_3 \qquad (1)$$

where q and p' relate to the octahedral (shearing) and hydrostatic (volumetric) invariants of the soil stress tensor, respectively; $\delta\varepsilon_g^p$ and $\delta\varepsilon_q^p$ are shearing and volumetric plastic strain increments, respectively. The critical state stress ratio M ($= q/p'$) mostly relates to the inter-particle frictional component of soil strength. Ignoring for a moment the f_1, f_2 and f_3 terms shown on the right side (which will be discussed next), Equation 1 shows that, for a saturated, disturbed soil to be in equilibrium, the work done by external loads applied to the soil (left side of Eq. 1) must be counterbalanced by the internal work mobilized through inter-particle friction at constant volume (right side of Eq. 1).

2.2 *Behaviour of stabilised rammed earth*

While stabilised rammed earth features most (or all) of the four issues described in Section 1 and is more complex than saturated, disturbed soil, critical state soil mechanics still provides a rigorous baseline for the expected mechanical response of any geomaterial. For a saturated geomaterial showing the first three features mentioned in Section 1, the right side of Eq. 1 requires additional terms to quantify changes in behaviour imparted by each additional factor (i.e., f_1, f_2 and f_3 would quantify the effect of transitional soil fabric, cementation and fiber reinforcement, respectively). While rigorous mathematical descriptions of functions f_1, f_2 and f_3 are out of the scope of this paper and constitute the focus of advanced research in geomechanics, a qualitative discussion on rigorous soil stabilisation methods is presented next that may be beneficial for stabilising rammed earth with or without waste materials.

3 INSIGHTS FROM SOIL STABILISATION

3.1 *Fabric-based mixture design*

Rammed earth consists of a mixture of soils with different particle sizes. From soil stabilisation, it is well known that the combination of various soil fractions (even if uniform individual soil fractions are used) leads to a mixture whose fabric is different than the fabric of any of the individual fractions. Figure 1 shows an idealised mixture obtained by combining three distinct uniform 'soils.' Two gaps can be observed in the particle size distribution of the resulting mixture, which is clearly less uniform than soils 1, 2 or 3. Figure 1 was obtained synthetically: all soil particles were assumed to have same shape and mineralogy, and arbitrary (weight-based) percentages were assigned to each soil fraction (25, 25 and 50 for soils 1, 2 and 3, respectively). The densest packing possible for soil 3 along with a simple analysis of the approximate number of particles in the resulting mixture (150, 15 and 3 for soils 1, 2 and 3, respectively) suggests perfect placement of finer particles from soils 1 and 2 within the voids formed between the larger particles of soil 3 (in its densest state) is not possible. The resulting mixture has a floating fabric (Salgado et al. 2000) even in its densest state. As a result, features of the mechanical behaviour of the mixture such as stiffness and strength, for example, cannot be predicted based on the idealized, expected behaviour of the predominant soil fraction (soil 3) alone. In this example, soils 1 and 2 will invariably push soil 3 particles apart and lower the stiffness and strength of the mixture—compared to other mixtures that could have been designed, for example, to maximize the filling of void spaces between soil 3 particles in their densest packing (Fig. 1) with particles from soils 1 and 2. This analysis can also be applied to real, compacted soil mixtures for an initial assessment of their fabric. As a minimum, Figure 1 can help assess whether a floating or nonfloating fabric would prevail in the resulting mixture.

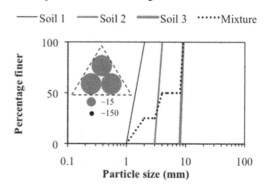

Figure 1. Mixture design based on particle size analysis.

Waste materials can also be analysed using this approach once their basic characteristics (e.g. specific gravity, particle shape, particle size distribution, etc.) are known. Such analysis can be particularly useful for preliminary selection of waste materials as their characteristics might lie outside values expected for typical soils. For example, class-F fly ash particles might look like perfect spheres and yet be hollow inside, thus lightweight. Granulated rubber from scrap tires has specific gravity slightly above unity whereas values for iron ore waste are typically on the upper end of spectrum for natural soils. Waste glass particles may be more angular than many natural sands. Such deviations in basic characteristics of a prospective waste material and their impact on the fabric of the resulting alternative soil mixture might be either detrimental or advantageous, depending on the intended use of the mixture (e.g. hollow particles can improve thermal insulation, angular particles can yield higher strength, etc.). Despite its synthetic nature, such basic fabric analysis can still provide insights into the characterisation and selection of waste materials being considered for the development of alternative soil mixtures.

3.2 Stabilizer design

The use of a rational soil stabilisation framework is desirable for the improvement or stabilisation of any soil mixture including rammed earth. This primarily involves deciding how much of a given stabiliser (e.g. hydrated lime, Portland cement, etc.) can be used to improve or stabilise a soil (or soil mixture) following fundamental soil mechanics principles. Soils of different composition and mineralogy require different types and amounts of stabilisers. For example, uniform quartz sands respond well to Portland cement stabilisation as inter-particle cementation stiffens and strengthens soil fabric (i.e., imagine small particles of soil 1 shown in Figure 1 representing cementitious compounds bonding the larger particles of soils 2 and 3, as opposed to just inert fillers). Similarly, hydrated lime is known to successfully improve or stabilise clayey soils as clay mineral double layers can be manipulated by using the right type and amount of lime stabiliser following basic soil plasticity analysis (Consoli et al. 2001). Such an approach has been successfully used for rammed earth design (Ciancio et al. 2014). The same rationale can be used to replace commercial stabilisers such as hydrated lime or Portland cement with waste materials with potential to yield similar levels of improvement. Examples abound in the literature on the successful use of waste materials such as carbide lime and fly ash, among others, to stabilise sandy and clayey soils (Consoli et al. 2001, Wiechert et al. 2011).

3.3 Design of discrete reinforcing elements

Fibers are successfully used to alter the mechanical response of engineering materials (Hull and Clyne 1996). The design concept is based on the volume-averaged sharing of external loading so that the composite is in equilibrium. For example, the external stress applied (σ_A) to a composite is internally balanced by individual fractions through

$$f\sigma_m + (1-f)\sigma_f = \sigma_A \tag{2}$$

where σ_m and σ_f are volume-averaged matrix and fibre stresses, respectively, and f is the volume fraction of reinforcement. Fiber reinforcement has also been used for soil stabilisation and adopted in recent systematic studies (Consoli et al. 1998). A similar methodology has been applied to reinforce a cemented geomaterial with waste fibers (van de Lindt et al. 2008). Fibers bridge cracks and increase tensile strength of cemented soils (Fig. 2). Employing such an approach to improve or stabilise rammed earth would be a natural extension of this technology.

3.4 Insights on saturation

Full saturation is assumed in most geotechnical analyses typically leading to conservative design parameters. For cemented soils (e.g. stabilised rammed earth), the mixture water content can also have a major impact on the chemical stabilisation process. Thus, saturation effects must be properly assessed not only during testing of stabilised rammed earth elements (e.g. triaxial testing of saturated specimens, unconfined compression of unsaturated specimens, etc.) but, perhaps more importantly, during the evolution of the stabilisation process. As an example, the saturated strength of a properly cured, stabilised rammed earth mixture may be higher than the unsaturated

Figure 2. Artificially cemented soil stabilized with recycled fibers from scrap tires.

strength of a similar mixture that is improperly cured (e.g. lack of water required for pozzolanic reactions can negatively affect calcium-based stabilization of soils and rammed earth).

4 POTENTIAL APPLICATIONS

4.1 *Stiffness and/or strength improvement*

Changes in soil stiffness and strength can be imparted by all methods discussed in Section 3. Macro- and micro-structural design of engineering materials is a powerful tool that tends to be overlooked in civil engineering applications. Such an approach may be particularly relevant for compacted soil mixtures such as rammed earth.

Well-graded soils can be stiffer and stronger than their uniform or gap-graded counterparts (assuming all other relevant state variables such as density and stress are properly accounted for). Similar improvements can be obtained by inducing artificial cementation to a soil, regardless of the type of cementitious material used. On the other hand, addition of discrete inclusions to a soil matrix can alter soil stiffness in a way that is related to the stiffness of the inclusion (compared to the stiffness of the original matrix material). Thus, the three methodologies described in section 3 can be employed (individually or in combination) to alter the stiffness and strength of rammed earth. Their use has also been validated in cases where waste materials were used in lieu of commercial additives and/or admixtures (Consoli et al. 2001, Wiechert et al. 2011, Dunham and Carraro 2014).

4.2 *Swell potential mitigation*

For soils with fine-grained fraction sensitive to changes in water content (e.g. expansive soils containing smectite clay minerals) or chemical composition (e.g. high-sulfate soils), a rigorous understanding of soil stabilisation principles is useful. If waste materials are considered either as inert fillers or as stabilisation admixtures, such understanding is critical as waste materials can widen the spectrum of chemical issues associated with the intended stabilisation process. Nevertheless, various systematic studies have shown that scrap tire rubber (Seda et al. 2007), fly ash (Wiechert et al. 2011), waste glass, waste fibers and other waste materials can be used to mitigate the deleterious volume change response of soils otherwise prone to large swell-shrink potential. If the fine-grained soil fraction of rammed earth contains smectite, any of the alternative materials mentioned above might help mitigate its swell potential.

4.3 *Thermal conductivity changes*

If properly engineered, mixtures of soils and waste materials can have thermal performance that is as good as the thermal performance of similar mixtures obtained with commercial materials. As for all other applications described previously, the key design aspect relates to being able to not only understand the thermal properties of individual materials but also of the final mixture. Just by knowing basic features related to their composition, waste materials such as rubber or fly ash can be expected to show lower thermal conductivity than typical soils, albeit for different physical reasons. Another example of a successful use of a mixture consisting of waste fiber from scrap tires and class-C fly ash was presented by van de Lindt et al. (2008) who used such a mixture to manufacture dry wall panels for house insulation. Replacement of specific particle size fractions of rammed earth with alternative materials such as scrap tire fiber, fly ash, waste glass, etc. can be demonstrated if proper engineering approaches such as those described and cited in this paper are systematically employed in comparative analyses.

5 CONCLUSIONS

Rammed earth is a man-made compacted mixture of soils whose characteristics can be controlled. There is great potential for rammed earth design to incorporate underlying concepts that rely on soil mechanics principles and that are environmentally friendly.

Basic soil characteristics (e.g. specific gravity, particle shape, particle size distribution, plasticity and simple volumetric indices) can provide useful information for the preliminary selection of waste materials for rammed earth stabilisation.

With minimal modifications (if any), rational soil stabilisation methods discussed in this paper may be used to systematically assess inherent features of saturated rammed earth such as transitional soil behavior, cementation and fiber-reinforcement. Recent studies on the beneficial use of waste materials for soil stabilisation may be relevant to rammed earth applications. Such studies may provide a starting point for rammed earth design with waste materials.

REFERENCES

Ciancio, D., Beckett, C.T.S. and Carraro, J.A.H. (2014) Optimum lime content identification for lime-stabilised rammed earth. Construction & Building Materials, 53(2014), 59–65.

Consoli, N., Prietto, P. and Ulbrich, L. (1998). Influence of Fiber and Cement Addition on Behavior of Sandy Soil. J. Geotech. Geoenviron. Eng., 124(12), 1211–1214.

Consoli N.C., Prietto P., Carraro J.A.H. and Heineck K. (2001). Behavior of compacted soil-fly ash carbide lime mixtures. J Geotech Geoenviron Eng. 127(9), 774–782.

Dunham-Friel, J. and Carraro, J.A.H. (2014). Effects of compaction effort, inclusion stiffness and rubber size on the shear strength and stiffness of expansive soil-rubber (ESR) mixtures. In: Geo-Congress 2014, ASCE, doi: 10.1061/9780784413272.352.

Fredlund, D.G. and Rahardjo, H. (1993) Soil mechanics for unsaturated soils. Wiley.

Hull, D. and Clyne, T.W. (1996). An introduction to composite materials. Cambridge, Cambridge University Press.

Martins, F., Bressani, L.A., Coop, M.R. and Bica, V.D. (2001). Some aspects of the compressibility behaviour of a clayey sand. Canadian Geotechnical Journal 38(6), 1177–1186.

Mitchell, J.K. and Soga, K. (2005). Fundamentals of soil behavior, Wiley, New York.

Salgado, R., Bandini, P. and Karim, A. (2000). Shear strength and stiffness of silty sand. J. Geotech. Geoenviron. Eng., 126(5), 451–462.

Schofield, A. and Wroth, P. (1968) Critical state soil mechanics. London, McGraw Hill.

Seda, J.H., Lee, J.C. and Carraro, J.A.H. (2007). Beneficial use of waste tire rubber for swelling potential mitigation in expansive soils. In: Geo-Denver Conference. ASCE, GSP No. 172, 1–9.

van de Lindt, J.W., Carraro, J.A.H., Heyliger, P.R. and Choi, C. (2008). Application and feasibility of coal fly ash and scrap tire fiber as wood wall insulation supplements in residential buildings. Resources, Conservation and Recycling. 52(10), 1235–1240.

Wiechert, E.P., Dunham-Friel, J. and Carraro, J.A.H. (2011). Stiffness Improvement of Expansive Soil-Rubber Mixtures with Off-Specification Fly Ash. In: Geo-Frontiers 2011, ASCE, pp. 4489–4497. doi: 10.1061/41165(397)459.

Rammed Earth Construction – Ciancio & Beckett (Eds)
© 2015 Taylor & Francis Group, London, ISBN 978-1-138-02770-1

A procedure to measure the in-situ hygrothermal behavior of unstabilised rammed earth walls

P.A. Chabriac, A. Fabbri, J.-C. Morel & J. Blanc-Gonnet
LGCB, CNRS-LTDS, UMR 5513, Ecole Nationale des Travaux Publics de l'Etat, Vaulx-en-Velin, France

ABSTRACT: Rammed earth is a sustainable material with low embodied energy. However, its development as a building material requires a better evaluation of its moisture-thermal buffering abilities and its mechanical behavior. Both of them are known to strongly depend on the amount of water contained in material pores and its evolution. Then the aim of this paper is to present a procedure to measure this key parameter in rammed earth walls by using two types of probes operating on the Time Domain Reflectometry (TDR) principle. A calibration procedure for the probes requiring solely four parameters is described. This calibration procedure is then used to monitor the hygrothermal behavior of a rammed earth wall ($1.5 \times 1 \times 0.5$ m), instrumented by five probes during its manufacturing, and submitted to insulated, natural convection and forced convection conditions. These measurements underline the robustness of the calibration procedure for a large range of water content, even if the wall is submitted to quite important temperature variations. They also emphasize the importance of gravity on water content heterogeneity when the degree of saturation is high, and the role of liquid-to-vapor phase change on the thermal behavior.

1 INTRODUCTION

Rammed earth is now the focus of scientific research though it is an ancient technic. Firstly, the modern earthen construction is sustainable because it is a natural and local material (Habert et al. 2013). Secondly the heritage of rammed earth buildings in Europe and in the world is still important (Fodde 2009). Maintaining this heritage needs scientific knowledge to apply appropriate renovations (Hamard et al. 2013). Modern rammed earth walls are load bearing structures and usually do not need insulation. Therefore several research studies have recently been conducted to study a wide range of characteristics of rammed earth: durability and sensitivity to water (Bui and Morel 2014, Bui et al. 2014); compressive mechanical characteristics (Bui et al. 2009); dynamic behaviour (Bui et al. 2011); capacity subject to lateral wind force (Cianco and Augarde 2013). Rammed earth wall houses are well known for their comfort of living. Indeed, earthen materials in general are materials which present the ability of regulating the temperature and relative humidity inside a building (Hall and Allinson 2010, Mc Gregor 2014). Measurement of *in-situ* soil water content is commonly carried out using TDR (Time Domain

Reflectometry) probes. These sensors are based on the measurement of soil dielectric constant, or dielectric permittivity (80 for pure water at 20°C (Lide 2001) and up to 14 for clay minerals (Fabbri et al. 2006)). So any change in the water content induces a change in its overall dielectric constant ε'_r (Cosenza & Tabbagh 2004). Several authors have proposed universal relations to link the dielectric constant to water content in different type of soils (e.g. Topp et al. 1980). These relations are good in first approximation but not enough accurate at low saturation. In addition they are only valid at a given temperature which is problematic in the case of on-site measurement. A TDR probe does not directly evaluate the dielectric constant but the travel time (τ) for the reflection of an electromagnetic wave between conductive rods of length (L). ε'_r is then written:

$$\varepsilon'_r = \left(\frac{c_0 \tau}{2L}\right)^2 - \left(\frac{\sigma}{\varepsilon_0 \omega}\right)^2 \left(\frac{L}{c_0 \tau}\right)^2 \qquad (1)$$

where c_0 is the speed of light in vacuum (2.9979×10^8 m · s⁻¹), σ the electrical conductivity (S · m⁻¹) of the material, ω the angular frequency (s⁻¹) and ε_0 the permittivity of free space (8.85×10^{-12} F · m⁻¹).

2 SENSORS

2.1 CS616

The CS616 probe has been developed by (Campbell & Anderson 1998) and is manufactured by Campbell Scientific. It is widely used in soil physics for its robustness, ease of use and low price. This probe does not allow access to the electrical conductivity of the material or to the angular frequency. As a result, this is not the "true" permittivity which is evaluated but an apparent permittivity ε_a' (Černy 2009) defined as:

$$\varepsilon_a' = \left(\left(\frac{\tau_m}{S_f} - 2t_d \right) \frac{c_0}{4L} \right)^2 \tag{2}$$

In this equation, according to the study of Kelleners et al. on this probe, L is the effective length of the rods and is equal to 0.26 m (Kelleners et al. 2005).

According to the specifications the sensor-to-sensor variability is ±0.015 $m^3 \cdot m^{-3}$ in saturated soil, the precision 0.0005 $m^3 \cdot m^{-3}$ VWC (Volumetric Water Content) and the accuracy ±0.025 $m^3 \cdot m^{-3}$ VWC for $\sigma < 0.5$ dS $\cdot$ m^{-1} and dry density of 1.55.

2.2 CS650

The CS650 sensor works on the same principle and is physically identical to the CS616. It converts the analogic signal into a digital signal and sends it to the datalogger via a SDI12 protocol. It is able to measure the temperature. It also evaluates the electrical conductivity and the angular frequency. Thus the "true" dielectric permittivity is measured according to equation (1). The specifications of the probe state that the precision in the permittivity measurement is ±2% + 0.6 in the range of 0 to 40.

2.3 Calibration procedure

The studied rammed earth comes from the construction site of a house built in 2011 in France. The earth is mixed with 2.5% (dry weight) of NHL5 lime. Characteristics of the earth are given in Table 1.

Calibration was made under laboratory conditions on two blocks containing a CS616 for the first one and CS650 for the second one manufactured by the same mason who built the house with a pneumatic rammer. Blocks dimensions are $45 \times 15 \times 9$ cm with dry density of 1.73 and gravimetric water content $w = 20.1\%$ for the block with the CS616 and $w = 19.6\%$ for the block containing the CS650. The followed calibration procedure is fully described elsewhere (Chabriac et al. 2014). It is assumed that the dielectric permittivity is

Table 1. Characteristics of the rammed earth.

Property	Symbol	Value
Dry density (–)	d_d	1.73
Porosity ($m^3 \cdot m^{-3}$)	ϕ	0.347
Dry electrical permittivity (–)	ε_r'	2.5
Dry thermal conductivity (W $\cdot$ m^{-1} $\cdot$ K^{-1})	λ	0.6
Water vapor resistance (–)	μ	10
Clay content (%)	–	16

defined as a bilinear function of volumetric water content ($\theta = w \cdot d_d$) and temperature (T). Thus the calibration equations are:

$$\varepsilon_r' = AT\theta + BT + C\theta + D; \quad \text{for CS650} \tag{3}$$

$$\varepsilon_a' = A'T\theta + B'T + C'\theta + D'; \quad \text{for CS616} \tag{4}$$

where A, B, C, D and A', B', C', D' are the calibration coefficients. The coefficients can be determined directly through the evolution of the dielectric permittivity with temperature at given water content. Four points for each probe are then needed to solve these equations instead of twelve as suggested by the manufacturer. The resulting coefficients are: $A = 0.84 \pm 0.07$; $B = 0.014 \pm 0.001$; $C = 47.07 \pm 1.79$; $D = 2.41 \pm 0.01$; for CS650 and to $A' = 1.38 \pm 0.18$; $B' = 0.014 \pm 0.001$; $C' = 51.06 \pm 4.69$; $D' = 2.40 \pm 0.01$ for CS616.

3 TESTS AT WALL SCALE DURING ITS DRYING

3.1 Description of the system

Five of these sensors have been inserted in a rammed earth wall ($1 \times 1.5 \times 0.5$ m) during its manufacturing. The wall is manufactured with the same earth used for calibration and by the same mason who built the house at gravimetric water content $w = 18.9\%$. Each layer is approximately 8 cm in height. Three CS616 are placed in the middle of the wall (0.25 m in depth) at three different heights: 0.1 m (bot.), 0.5 m (mid.) and 0.9 m (top) Figure 1.

At 0.5 m in height, a CS616 is placed at 0.1 m of the left surface and a CS650 at 0.1 m of the right surface. At this height, sensors are placed in a staggered configuration to avoid electromagnetic interferences. Temperature is measured using band-gap sensors placed in the layer immediately above the one containing the TDR probes. The wall is placed in a cork insulated (0.1 m) box developed in the laboratory with two doors on each side of the wall allowing it to dry. Two band-gap sensors are also installed in the ambiances. Implementation of the sensors in the wall is described on Figure 1. For

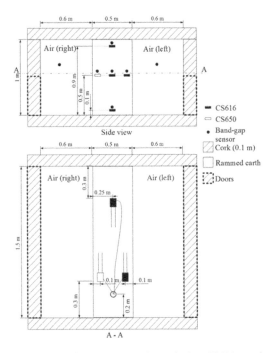

Figure 1. Side and top view of the CS616s and CS650 installed in a rammed earth wall during its manufacturing.

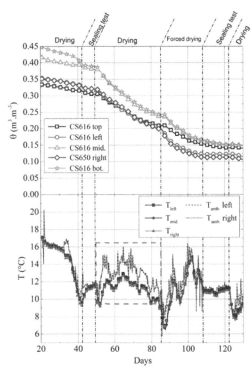

Figure 2. Evolution of the VWC (θ) and temperature (T) in the wall during its drying for 130 days.

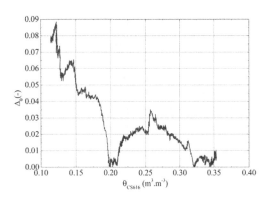

Figure 3. Deviation on the VWC measured by the CS616 and the CS650 on the left and right in the wall at 0.5 m high.

technical reasons the wall was protected by a plastic film to slow its drying after its manufacturing for 22 days. Data are then recorded for 130 days at a frequency of one point per hour. Two types of drying are studied: natural when doors are open (from the 22nd to 44th, 51st to 87th and 124th to 130th day) and forced with the doors open and an air circulation performed by a fan on each side of the wall (from the 87th to 110th day). These drying sequences are interspersed by sealing tests of the box (from the 44th to 51st and 110th to 124th day). Recorded data are given on Figure 2.

3.2 Accuracy of the calibration

Hydric and thermal conditions were identical on both side of the wall during its drying. So, the water content measured by the CS616 and the CS650 (respectively on the left and right side) at 0.5 m in the wall should provide the same values. The deviation between the probes is calculated with:

$$\Delta\theta = \frac{|\theta_{cs616} - \theta_{cs650}|}{\theta_{cs616}} \qquad (5)$$

The evolution of this deviation is shown on Figure 3. The average deviation is 3.3% with a peak value of 9% for $\theta = 12.4\%$). Moreover, the wall is submitted to temperature variations (from 7

to 20°C) during these 130 days (Figure 2) and the results are still close. This shows that the calibration is accurate even with temperature variations.

3.3 Influence of gravity

The effect of gravity on water can also be observed during the drying of the wall. A decrease of the water content with height is measured following $\theta_{bot.} > \theta_{mid.} > \theta_{top}$. The difference in water content

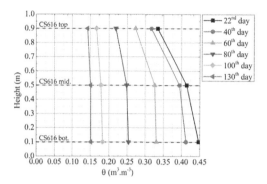

Figure 4. Volumetric water content as a function of the height in the wall during its drying.

decreases with time (Figure 4) to become equal over the height as capillary suction becomes preponderant over gravity effect.

This result shows that gravity effect should be taken into account at least when earth building are submitted to high moisture level (>10% gravimetric water content). This happens after the manufacture of the wall, at early ages (few months) or when the wall of heritage building is submitted to pathologies such as capillarity rises or water infiltration.

3.4 *Hygrothermal behavior*

On Figure 2 it is also possible to see that the temperatures in the right and left ambiances are significantly higher than the temperature measured in the wall ($T_{mid.}$) during natural drying sequence (framed from 50th to 85th day).

During this sequence the volumetric water content measured by the CS616 in the middle of wall decreases from 30% to 20%. During this sequence the temperature in the wall is lower by approximately 1.5°C in average than in the ambiances. This shows that the evaporation of the water contained in the wall is absorbing energy and naturally refresh the wall.

4 CONCLUSION

A procedure to measure the hygrothermal behavior of rammed earth wall is presented using commercial TDR and band-gap sensors. A wall is implemented with five TDR and band-gap sensors during its manufacturing. A simple calibration procedure is validated during the drying of the wall. The main advantage of this calibration is that it requires only four parameters instead of twelve. The effect of gravity on water has also been emphasized. This effect disappears as soon

as capillary forces become predominant over gravity in the wall pores. The gravity should be taken into account when modelling at least at early ages of the wall or when it is submitted to pathologies. The hygrothermal behavior of a rammed earth wall is also presented. The evaporation of water lowers the temperature in the wall and refreshes it naturally.

ACKNOWLEDGEMENT

The authors wish to thank the French national research agency ANR (PRIMATERRE—ANR-12-VBDU-0001-01 Villes et Bâtiments durables) for the funding of this project.

REFERENCES

Bui, Q.B., Morel, J.C., Hans, S., Meunier, N. 2009. Compression behaviour of nonindustrial materials in civil engineering by three scale experiments: the case of rammed earth, *Materials and Structures* 42(8): 1101–1116.

Bui, Q.B., Hans, S., Morel, J.C., Do, A.p., 2011. "First exploratory study on dynamic characteristics of rammed earth buildings", *Engineering Structures*, 33: 3690–3695.

Bui, Q.B., Morel, J.C., Hans, S., Walker, P. 2014. Effect of moisture content on the mechanical characteristics of rammed earth, *Construction & Building Materials* 54: 163–169.

Bui, Q.B. and Morel, J.-C. 2014. First exploratory study on the ageing of rammed earth material, *Materials: to be published.*

Campbell, G.S., and Anderson, R.Y. 1998. "Evaluation of Simple Transmission Line Oscillators for Soil Moisture Measurement." *Computers and Electronics in Agriculture* 20(1), 31–44.

Černy, R. 2009. "Time-Domain Reflectometry Method and Its Application for Measuring Moisture Content in Porous Materials: A Review." *Measurement* 42(3): 329–336.

Chabriac, P.A., Fabbri, F., Morel, J.C., Laurent, J.P., Balnc-Gonnet, J. 2014. "A Procedure to Measure the in-Situ Hygrothermal Behavior of Earth Walls." *Materials* 7: 3002–3020.

Ciancio, D., Augarde, C. 2013. "Capacity of unreinforced rammed earth walls subject to lateral wind force: elastic analysis versus ultimate strength analysis", *Materials and Structures* 46: 1569–1585.

Cosenza, Ph., Tabbagh, A. 2004. "Electromagnetic Determination of Clay Water Content: Role of the Microporosity." *Applied Clay Science* 26: 21–36.

Fabbri, A., Fen-Chong T., Coussy O. 2006. "Dielectric Capacity, Liquid Water Content, and Pore Structure of Thawing–freezing Materials." *Cold Regions Science and Technology* 44(1): 52–66.

Fodde, E. 2009. Traditional earthen building techniques in Central Asia, *Journal of Architectural Heritage* 3:145–168.

Habert, G., Castillo, E., Vincens, E., Morel, J.C., 2013. Power: a new paradigm for energy use in sustainable construction, *Ecological Indicators* 23: 109–115.

Hall, M.R., Allinson, D. 2010. Transient numerical and physical modelling of temperature profile evolution in stabilised rammed earth walls, *Applied Thermal Engineering* 30(5): 433–441.

Hamard, E., Morel, JC., 2013. A procedure to assess suitability of plaster to protect vernacular earthen architecture, *Journal of Cultural Heritage* 14: 109–115.

Kelleners, T.J., Seyfried, M.S., Blonquist, J.M., Bilskie, J., Chandler, D.G. 2005. "Improved Interpretation of Water Content Reflectometer Measurements in Soils." *Soil Science Society of America Journal* 69 (6): 1684.

Lide, D.R. 2001. *Handbook of Chemistry and Physics 2001–2002 - CRC Press - 82nd Edition*.

McGregor, F., Heath, A., Fodde, E. Shea, A. 2014. Conditions affecting the moisture buffering measurement performed on compressed earth blocks. *Building and Environment* 75: 11–18.

Topp, G.C., Davis, J.L., Annan, A.P. 1980. "Electromagnetic Determination of Soil Water Content: Measurements in Coaxial Transmission Lines." *Water Ressources Research* 16(3): 574–582.

Rammed Earth Construction – Ciancio & Beckett (Eds)
© 2015 Taylor & Francis Group, London, ISBN 978-1-138-02770-1

Notched mini round determinate panel test to calculate tensile strength and fracture energy of fibre reinforced Cement-Stabilised Rammed Earth

D. Ciancio & C. Beckett
The University of Western Australia, Crawley, Australia

N. Buratti & C. Mazzotti
University of Bologna, Bologna, Italy

ABSTRACT: The use of natural fibres (like hemp or bamboo) to improve the mechanical performances of rammed earth structures is not new in construction practice in many parts of the world. However, little scientific investigation has been carried out so far to better understand the real improvement obtained by the addition of fibres. In a recent publication [1], the feasibility of notched mini Round Determinate Panels (mRDP) has been investigated with the aim of deriving a procedure to estimate the intrinsic material properties of Fibre Reinforced Shotcrete (FRS). It was found that it was possible to recover the tensile strength and the fracture energy of the material using an inverse analysis of the experimental data and the well-known Olesen constitutive model [2]. In this paper, the use of the notched mini round determinate panel test to characterise the post-cracking performances of Cement Stabilised Rammed Earth (CSRE) was investigated. For quality control issues, in this study the soil mix consisted of crushed limestone stabilised with 8% cement by soil mass and compacted at its optimum water content (11%). Three specimens were made of CSRE alone and three samples were made of fibre-reinforced CSRE. The fibres used in this experimental campaign were unbundled synthetic copolymer fibres, 54 mm long and 0.3 mm thick. This paper discusses the applicability of a laboratory test conceived for concrete samples to rammed earth specimens. It also presents the comparison between the performances of CSRE materials with and without fibres.

1 INTRODUCTION

The addition of fibres (steel or synthetic) to concrete mixes is a well-established practice all over the world. It is well understood that the presence of fibres increases the post-cracking performance of concrete by controlling crack opening and improving load distribution. Natural fibres have been used in traditional (not cement-stabilised) rammed earth construction for a long time. Their function is however still not clear. There are two key arguments: i) that natural fibres help to control the opening of cracks; ii) that natural fibres help to create channels on the surface of rammed earth walls along which rain water flows. In this way, the erosion is localised and it is easier to repair the wall.

The addition of fibres to concrete is straightforward: a proper fluid state of the mix at casting guarantees a regular distribution. Furthermore, fibres orientation can be adjusted during casting. Rammed earth mixes are quite dry; the addition of fibres requires special care to produce a uniform spreading. Furthermore, the compaction process might bend or damage some fibres. If the fibre reinforced rammed earth mix is not properly put together and compacted, fibres might actually lead to a decrease of the mechanical performance of the material.

The aim of this work is to provide a preliminary understanding of the effects of macro-synthetic fibre addition on the post-cracking performance of SCRE. The mRDP test is used here, as opposed to other traditional tests (like flexural or Brazilian tests), to evaluate CSRE fracture energy, an essential component in modelling CSRE post-cracking behaviour. Results will also be used to determine the suitability of the mRPD test for use with rammed earth materials.

2 MATERIAL AND PROCEDURE

2.1 *Material*

Crushed limestone was used in this work in the place of natural soil due to its improved quality control, ready availability and preferential use for CSRE construction in Perth, WA. Oven-dried crushed limestone (particle size distribution shown in Figure 1) was combined with 8% cement by mass (again typical of CSRE construction in Perth).

For tests conducted with fibres, 0.2% by volume macro-synthetic fibres was added to the dry material mix. The fibres used in this investigation (Figure 2) were macro-synthetic polypropylene (54 mm long, 0.3 mm diameter, aspect ratio = 180, tensile strength 680 MPa) designed to increase the load transfer and post-crack performance in concrete mixes. Each bundle of fibres was broken down to individual strands to help spread fibres evenly during mixing. The relation of the dosage to material volume, rather than mass, is necessary due to the construction practise of preparing rammed earth mixes by known volumes.

2.2 Procedure

Specimens of size Ø540 mm, 75 mm depth were manufactured for use with the mRDP test as shown in Figure 3. This size was selected as being of sufficient size to be used with existing testing equipment whilst being light enough to be handled manually. Material was compacted at its Optimum Water Content (OWC) of 11.1% (for both mix types), found using the modified Proctor compaction test (AS 1289.5.2.1 [3]) for CSRE mixes with and without fibres.

Specimens comprise a single compaction layer to prevent delamination during testing. Material was first compacted using a manual tamper, as shown

in Figure 3, to ensure that it remained within the mould. An electric Bosch GHS 11 E jackhammer with a 100 mm diameter round steel end plate was then used to compact the material to a thickness of 75 mm. Finally, a light hand tamper was used to create a level surface, as shown in Figure 4. Specimens were then left to cure for 24 hours under moist hessian sacking prior to being removed from the mould and cured for an additional 27 days at $94 \pm 2\%$ relative humidity and $21 \pm 1°C$.

After curing for 14 days, three radial notches, symmetrically trisecting the specimen (i.e. at 120 degrees), were cut into the specimen underside; as this surface in contact with the mould baseplate it provided the smoothest surface for testing. The notch depth was of 14 mm.

At 28 days, samples were tested by placing them on three symmetric point supports and by applying a point load at the centre, at a rate of 4 mm/min. The load and the central vertical displacements were recorded until failure occurred. Furthermore, one micro-yoke per notch (3 per sample) was installed as

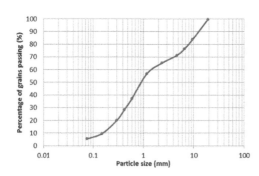

Figure 1. Particle size distribution of rammed earth mix.

Figure 2. Macro-synthetic fibres used. On the left: bundles; on the right: individual strands.

Figure 3. Initial manual compaction of specimen.

Figure 4. Levelled surface.

76

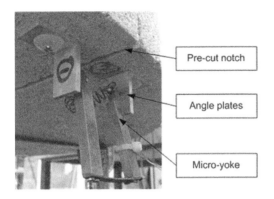

Figure 5. Positioning of micro-yoke and angle plates.

shown in Figure 5 in order to record Crack Mouth Opening Displacements (CMODs). For further explanations on the testing equipment, please refer to [1]. Moment per unit length-crack rotation angle curves for each specimen were then determined according to testing conducted as per [1].

3 RESULTS

Figures 6 and 7 show the moment per unit length m vs crack rotation angle θ curves, respectively for the unreinforced and fibre-reinforced RE samples. For each sample, 3 curves are reported, one per notch. The presence of the fibres does not seem to affect the peak value (around 1.6 kN mm/mm for both unreinforced and reinforced samples). The peak value of m is strictly related to the tensile strength properties of the material (for further details please refer to [1]). This means that the fibres do not alter the tensile strength properties of CSRE. This is similar to what observed in concrete.

However, the post peak performance seems quite different with and without fibres. For a crack rotation angle of 0.015 rad, the unreinforced samples show no residual strength while the fibre-reinforced samples show a residual strength of 12.5% the peak load.

The m-θ curves have been numerically obtained by inverse analysis using the constitutive model proposed by Olesen [2]. For further details on these type of analysis please refer to [1]. One of the mechanical parameters of the Olesen's model is the fracture energy. This parameter has been calibrated so that the moment per unit length vs. CMOD curve obtained via experimental procedure matches the numerical curve obtained using Olesen's model. The accuracy of this procedure is shown in Figure 8.

Table 1 reports the fracture energy values found in the inverse analysis. The fibre-reinforced samples

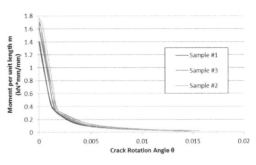

Figure 6. Moment per unit length-crack rotation angle curve for unreinforced RE samples.

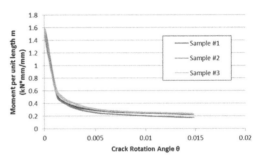

Figure 7. Moment per unit length-crack rotation angle for fibre-reinforced RE samples.

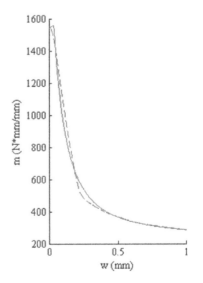

Figure 8. Numerical (dashed line) and experimental (continuous line) curves of moment per unit length m vs. crack mouth opening displacement w.

show fracture energy values 4.3 times higher than the samples without fibres. It is clear that this is due to the presence of the fibres, which have a beneficial effect on the load transfer during the cracking process. In

Table 1. Fracture energy values from inverse analysis.

| Sample # | Fibres | Fracture energy Gf [N/mm] | | | | |
		Crack 1	Crack 2	Crack 3	Sample average	Series average
1	no	0.058	0.055	0.056	0.056	
2	no	0.065	0.066	0.076	0.069	
3	no	0.070	0.063	0.063	0.065	0.063
1	yes	0.279	0.266	0.294	0.280	
2	yes	0.240	0.223	0.282	0.248	
3	yes	0.276	0.281	0.317	0.291	0.273

other words, a cracked structural element reinforced with fibres shows the ability to carry some loads when the same structural element without fibres would have no capacity at all. This is an interesting outcome that might open the discussion over the use of fibres for structural elements in earthquake areas, where the capacity of the material to dissipate energy is a crucial aspect of the seismic design.

Unfortunately, the research available in the literature on the fracture energy of rammed earth is very limited. To evaluate the reliability of the proposed experimental procedure, the values of fracture energy obtained in this work have been compared to the ones found for similar materials. Petersson [4] found that for limestone concrete beams, the fracture energy obtained with the three-point bend test was in the range of 0.055–0.06 N/mm. The CSRE mix used in this work does not substantially differ from a limestone concrete and the similarity between the values of Petersson and the ones shown in Table 1 confirms the goodness of the use of the mini round determinate panel test for the evaluation of the fracture energy.

Carnovale [5] estimated the fracture energy of macro-synthetic fibre reinforced concrete panels to be around 0.31 N/mm. This value is of the same order of magnitude of those found in this study and reported in Table 1.

As mentioned in the Introduction section, the addition of fibres not always has a favourable effect in the post-cracking mechanical properties of the mix. Corbin and Augarde [6] used wool fibres to reinforce cylindrical rammed earth samples tested using the wedge splitting test. For samples stabilised with 8% of cement, the fracture energy decreased from 0.022 N/mm for samples without any fibre, to 0.010 N/mm for samples containing 2% of fibre by mass.

4 CONCLUSIONS

This paper investigated the improvement in the post-cracking mechanical performance of cement-stabilised rammed earth samples obtained by the addition of macro-synthetic fibres in the material mix. The improvement was quantified by the measurement of the material fracture energy, experimentally obtained using the mini notched round determinate panel test. It was found that the addition of 2% by volume of synthetic fibres to the rammed earth mix increased the material fracture energy of 4 times, providing some ductility properties that are usually crucial in structural elements used in seismic areas.

The values found in this work were compared with others available in the literature for similar materials (plain limestone concrete and micro-synthetic fibre reinforced concrete). The similarity between these values validated the use of the experimental procedure implemented in the experimental campaign proposed in this research.

ACKNOWLEDGMENT

The first and second authors would like to thank the Australian Research Council and the Department of Housing for providing the funding necessary to complete this work through the ARC grant LP110100251.

REFERENCES

[1] Ciancio, D., Mazzotti, M. and Buratti, N. (2014) Evaluation of Fibre-Reinforced Concrete Fracture Energy through Tests on Notched Round Determinate Panels with Different Diameters, *Construction & Building Materials*, Vol. 52, pp. 86–95.
[2] Olesen, J.F. (2001) Fictitious Crack Propagation in Fibre-Reinforced Concrete Beams. *Journal of engineering Mechanics*, Vol 27(3), pp. 273–80.
[3] Standards Australia (2003). AS 1289 Methods of testing soils for engineering purposes. Method 5.2.1: Soil compaction and density tests—Determination of the dry density/moisture content relation of a soil using modified compactive effort.
[4] Petersson, P.E. (1980) Fracture energy of concrete: method of determination. *Cement and Concrete Research*, Vol 10, pp. 78–89.
[5] Carnovale, D.F. (2013) Behaviour and analysis of steel and macro-synthetic fibre reinforced concrete subjected to reversed cyclic loading: a pilot investigation, Master's Thesis at the University of Toronto.
[6] Corbin, A. and Augarde, C. (2014) Fracture energy of stabilised rammed earth, Procedia Materials Science, Vol 3, pp. 1675–1680.

Rammed Earth Construction – Ciancio & Beckett (Eds)
© *2015 Taylor & Francis Group, London, ISBN 978-1-138-02770-1*

Modular rammed earth masonry block

A.J. Dahmen
School of Architecture and Landscape Architecture, University of British Columbia, Canada

B.J.F. Muñoz
Watershed Materials LLC, Napa, CA, USA

ABSTRACT: This paper describes a modular rammed earth masonry block fabricated from soil materials that could radically improve the environmental profile of one of the most common construction materials on the planet. Although compressed earth blocks have been widely available for half a century, most are specialized building products that cannot be easily integrated with the conventional concrete masonry construction due to differences in form and performance. In contrast, the modular rammed earth block described in this paper is functionally interchangeable with ordinary concrete blocks and meets common performance specifications for concrete masonry. The block is designed to be interchangeable with conventional concrete block, and includes hollow cells to permit the placement of grout and reinforcing steel. The block reduces embodied energy by as much as fifty percent compared to conventional concrete masonry due to the reduction of energy-intensive Portland cement binders, dramatically reducing CO_2 emissions of this common construction material.

1 RAMMED EARTH ENTERS THE MAINSTREAM

During the summer of 2014, the cough drop giant Ricola quietly opened the Kräuterzentrum, a €16M herb processing facility sited in a picturesque meadow outside of Basel, Switzerland (Neue Zurcher Zeitung, 2014). The 3,330 m² building is a simple monolithic mass whose powerful expression comes primarily from the walls of the building, which are constructed of prefabricated sections of rammed earth.

The Kräuterzentrum might be read as evidence that rammed earth has finally entered the mainstream in the developed world after decades of niche popularity among architects interested in sustainability. Designed by Herzog and de Mueron, one of Europe's leading architecture firms, for a corporate client with annual earnings of €240M, it is at present the largest rammed earth building in Europe. However, despite high visibility projects and areas of local popularity, rammed earth remains a vanishingly small percentage of total construction in developed countries. This paper explores some of the reasons that this is the case, and one offers a response in the form of a new material that draws on these lessons.

1.1 *Rammed earth: History and global context*

Rammed earth is one of the world's oldest building technologies, and earth building traditions have persisted for thousands of years in various regions (Jaquin et al. 2008, Lowe, 2012). It has been speculated that one third to half of the work's population lives in unfired earthen housing (Avrami & Guillaud, 2008), much of this rammed earth. Its contemporary resurgence began in the developed world roughly fifty years ago, when a small but ardent number of independent innovators pioneered new installation techniques appropriate to the developed world. These efforts accompanied shifting attitudes about earth housing concurrent with growing interest in what came to be known as environmental sustainability. From these early efforts, it is now possible to point to significant rammed earth projects on six continents. The soils of Antarctica alone remain unrammed.

2 BENEFITS AND ADOPTION RATES OF RAMMED EARTH

2.1 *Benefits of rammed earth construction*

The benefits of building with earth are many, and more pronounced as our awareness of the effect of construction on the environment increases. Globally, the built fabric accounts for as much as forty percent of all energy use (Perez-Lombard et al. 2008) of which perhaps 15–20% is due to the energy embodied in materials (Dixit et al. 2010). Earthen construction methods, which use natural mineral subsoils with or without additional binders and amendments, can be used to create durable buildings with lower embodied energy than conventional masonry or cast in place concrete buildings (Reddy & Kumar, 2010). The environmental advantages of rammed earth come into sharpest focus when it is compared to concrete, the material it most often replaces. The production of cement and concrete accounts for 6–7% of CO_2 emissions worldwide, and each ton of cement manufactured releases approximately 900 kg of CO_2 into the atmosphere (Chaturvedi & Ochsendorf, 2004). In contrast, traditional rammed earth uses natural binders and soils sourced at or nearby the building site, lowering the energy profile even further. While unstabilized rammed earth does not have the compressive strength of concrete or fired masonry, with proper design it can be used to create safe, efficient and enduring structural systems that require only a small fraction of its available material strength.

2.2 *Adoption rates of rammed earth*

Despite these advantages, the adoption of rammed earth construction in contemporary mainstream construction has been limited relative to other construction methods. While no comprehensive global survey of rammed earth buildings in the developed world has been written, it is unlikely that more than 50,000 new rammed earth structures have been built in the developed countries since the early 1970s, when the method began to regain popularity. In reality, the number is probably closer to 10,000 structures. By comparison, during the 30 year period between 1980 and 2010, an average of 1.7 million new residential units were completed each year in the US alone (US Dept. of Energy, 2014). Comparing these figures suggest that rammed earth accounts for a vanishingly small percentage of overall construction, despite powerful environmental advantages and positive cultural associations in the developed world.

A full exploration of the complex cultural, technical, and administrative reasons preventing the widespread adoption of rammed earth is beyond the scope of this paper. However, three areas stand out as significant contributors: the high level of variability of site soils, which makes performance somewhat unpredictable, the difficulty of integrating monolithic rammed with existing construction methodologies (Hall & Swaney, 2012), and the cost of rammed earth, which has remained high in the developed world despite attempts at mechanization and prefabrication. Half a century into the global resurgence of rammed earth, with environmental concerns ever more pressing, it is worth asking the question: what other pathways to adoption of rammed earth might be explored to bring its considerable benefits to a wider audience?

3 MODULAR RAMMED EARTH

3.1 *Addressing the barriers to adoption*

One way to address the barriers to adoption facing rammed earth is to unitize it so that it can be produced under controlled conditions. This would enable rammed earth to consistently meet common performance standards and fit within established construction methodologies. This paper documents the development of Rammed Earth Masonry Units (REMUs) by Watershed Materials LLC, a California based masonry materials developer. The REMUs, sold under the trade name of Watershed Block, are functionally interchangeable with Concrete Masonry Units (CMUs). CMU is the technical term for concrete block, which is one of the most common construction materials on the planet. REMUs are hollow rectangular $200 \times 200 \times 400$ mm units identical in form to that of a conventional CMU. Since the REMU shares the same dimensions as common CMU, it can be easily integrated in current masonry construction activities to meets the steadily increasing demand of sustainable masonry materials in domestic and global markets.

3.2 *Modular earth masonry, past and present*

Modular earth masonry has a history stretching back 10,000 years. Adobe blocks probably predate rammed earth and were widely used where stone was unavailable, the clay of local soils was sufficient and there was ample sunshine for drying. More recently, Martin Rauch, one of the foremost rammed earth builders in Europe, has used prefabricated rammed earth on many of his projects in Europe to counteract the high cost of labor there (Kapfinger, 2002). However, the prefabricated blocks produced by Rauch are large enough to require a crane to set, requiring specialized knowledge and equipment, which limits their application.

The most common form of unitized earth masonry currently in the Compressed Earth Block (CEB). Manual and automated CEB equipment has been widely available since the 1950's (Morel et al. 2007). However, none of the CEB production machines produce blocks in the shape of a typical CMU, nor are they designed to produce blocks that meet the standards governing unitized masonry construction. As a result, most CEB machines are primarily marketed to "do-it yourselfer" owner-builders, and CEB production has for the most part remained limited to developing regions where codes and construction are more forgiving toward non-standard approaches.

3.3 *Modular rammed earth*

In contrast to traditional adobe blocks and CEBs, REMUs are designed to meet the compressive strength, water absorption and wet-dry durability of the ASTM C90 Standard Specification for Loadbearing Concrete Masonry Units and ASTM C426 Standard Specification for Linear Drying Shrinkage of Concrete Masonry Units, whose threshold values are summarized in Table 1.

The challenge is to produce REMUs that comply with these performance criteria with minimal quantities of energy-intensive OPC binder content. The REMU manufacturing process redesigns typical CMU production in order to limit or eliminate OPC stabilizers. The static compaction (steady pressure) typically used to produce CEBs as been combined with high frequency impact to maximize consolidation, achieving dry densities above 2400 kg/m³. The overall mechanical performance, in particular water absorption and freeze-thaw durability, of earth units under these compaction conditions is significant enhanced, and enables the company to meet the relevant ASTM standards with 4% OPC by weight, a significant reduction from the OPC used in a conventional CMU. This also represents a reduction of OPC stabilizers typical for rammed earth in developed world applications, where it is generally between 6–8% or higher (Treloar & Fay, 2001; Lax, 2010; Hall & Swaney, 2012).

3.4 Manufacturing rammed earth units

It is possible to produce REMUs in controlled settings using specific equipment designed to integrate with existing CMU manufacturing lines. This equipment can be used to replace conventional CMU production equipment while maintaining existing feeding, curing and other infrastructure wherever possible. Market studies have indicated that production lines must be capable of producing hollow-celled blocks at an average rate of 3 blocks per minute to compete with CMU pricing. The target cost for REMU production equipment capable of the output above is US$250 K, comparable to conventional CMU machines of similar output capacity.

3.5 Environmental advantages

The energy required to produce a product is referred to as embodied energy, which is closely correlated with embodied carbon, which measures the emissions of CO_2 to atmosphere. A CMU typically contains between 8–12% by weight. OPC, which is the most expensive (Dahmen, 2013a) and energy intensive (Marceau et al. 2007) ingredient in a typical block. 91% of the embodied carbon in a CMU is due to cement content, as summarized in Table 2. Currently, in-house mix designs of REMUs only have 4% by weight. of OPC. These OPC reductions conservatively translate into 50–60% reductions in CO_2 emissions during the manufacturing of REMU in comparison with common and architectural CMUs, respectively.

The environmental benefits of REMUs extend beyond reduced carbon emissions. REMUs reduce the energy and water required to process and wash aggregates, and decrease environmental damage caused by mining and transporting raw materials by expanding the range of suitable recycled aggregates and co-locating production.

3.6 Market reaction

Designing the REMU to compete with CMU dramatically increases the total available market from the niche markets occupied by monolithic rammed earth. According to industry research, the 2012 concrete block and brick market were valued at $4.3B, with forecasts projecting positive growth of 8.2% to $5.9B by 2016 (Anything Research 2012). In total, concrete block products account for $2B for common CMU and $552M for high grade architectural CMU.

Although no formal studies on market reaction to Watershed Blocks have been conducted to date, it is possible to characterize initial customer response to pilot production runs and a 300m² demonstration house completed in the Bay Area of San Francisco, California in 2013, and meetings with CMU producers on the West Coast and Southwestern regions of the United States. On aggregate, these initial reactions have suggested

Table 1. Property threshold values of relevant ASTM standards.

Property	Threshold value	ASTM
Dry Density	≥125 pcf (2000 kg/m³)	C90
Compressive Strength	≥1900 psi (13.1 MPa)	C90
Water Absorption	≤15%	C90
Liner Shrinkage	≤0.065%	C426

Table 2. Energy required for CMU production.

Category	Energy (GJ)	% of total
Cement	0.691	68
Aggregate	0.038	4
Transportation	0.059	6
Plant operations	0.227	22
Total	1.015	100

Note: The production of cement accounts for 68% of the total embodied energy but 91% of total CO_2 emissions primarily due to CO_2 released during the calcination phase of cement production. Source: Analysis of PCA study by Watershed Material.

that interest in REMUs is high from architects and their clients, as well as existing CMU manufacturers. As might be expected, these groups are motivated by different aspects of the product. More surprisingly, sustainability is often a secondary concern for both groups.

Architects and owners respond most to the aesthetics of the REMU, which offers rich appearance of rammed earth in a unitized form. The natural mineral pigments and aggregates give the block an appearance of sedimentary stone, which architects and their clients cite as superior to the appearance of cement-based blocks (Dahmen, 2013b). Its appeal is partly due to the subtle variation, and also because it projects an image of sustainability. Improved environmental performance, in the form of reduced embodied energy over CMU, appears to be a secondary, and less marketable quality to architects.

Environmental sustainability of the REMU is also often of secondary concern to CMU manufacturers. Instead, manufacturers see the potential of the REMU production equipment as primarily economical. REMUs contain approximately half the OPC by compared to conventional CMU, which is a significant reduction of the most expensive ingredient in a CMU. This concern generally outweighs the positive benefits of lower embodied energy of the REMU over CMU where manufacturers are concerned.

3.7 Future directions

The REMU currently developed by Watershed Materials LLC reduce are an incremental improvement over existing CMU, reducing embodied carbon by 50–60% and radically expanding the available market for rammed earth. The future development of rammed earth masonry will explore the use of alternative binders to completely eliminate the use of OPC. Recent research conducted at Watershed Materials has demonstrated the possibility of producing strong and durable masonry products through exclusive alkali activation of naturally-occurring soil aluminosilicates minerals, a chemical activation process known as geopolymerization. Demonstrating the feasibility of achieving geopolymerization in alkali-activated soils is significant because it represents a means of producing durable masonry materials using common mineral soil, a virtually inexhaustible resource. The energy required to produce the geopolymer blocks, which will be marketed under the trade name ZeroBlock,™ is estimated to be 80% lower than conventional CMU due to natural aluminosilicates used as binders.

4 CONCLUSION

Rammed earth, one of the oldest building technologies on the planet, has the potential to reduce the embodied carbon in construction over other more energy intensive materials. However, the difficulty of integrating rammed earth with mainstream construction practices in the developed world has resulted in high costs and prevented its widespread adoption in these regions. Despite these difficulties, rammed earth techniques have the potential to address growing environmental concerns with cement-based masonry materials and practices. The REMU applies rammed earth methods to address the significant CMU market. REMUs contain approximately 50–60% less embodied carbon than conventional CMU, primarily due to reduced cement content, but are functionally interchangeable in form and performance. Large scale production of the blocks is possible in factory setting. Future products will target geopolymerization as a way to eliminate OPC content entirely, resulting in a block with 80% lower embodied carbon than a conventional block.

REFERENCES

Anything Research 2012. *2012 Report on Concrete Block and Brick Manufacturing*, industry report.

Avrami E, Guillaud H. 2008. In: Avrami E, Guillaud H, Hardy M, eds. *Terra literature review—an overview of research in earthen architecture conservation.* Los Angeles (United States): The Getty Conservation Institute. p. xi.

Chaturvedi, S. and Ochsendorf, J. 2004. Global Environmental Impacts due to Cement and Steel. *Structural Engineering International* Vol 14 Issue 3: pp.198–200.

Dahmen, J. 2013a. Telephone correspondence with Dennis Ceolin, Regional Operations Manager, Basalite Concrete Products and Ricardo Birkner, Director of International Sales at Columbia Machine, Inc., a leading CMU equipment manufacturer.

Dahmen, J. 2013b. personal interviews with Brian Gerich and six other senior associates at Mahlum Architects, a prominent Seattle-area architecture firm, and 20 mid level and senior associates at Olsen Kundig, an award—winning architecture firm in Seattle, WA with approximately 100 employees. Similar reactions to product samples were observed in interviews conducted at architecture firms located in the San Francisco Bay Area (CA) and Las Vegas, NV.

Dixit, M. Fernández-Solís, J. Lavy, S. and Culp, C. 2010. Identification of parameters for embodied energy measurement: A literature review. *Energy and Buildings*, Vol. 42, Issue 8, pp. 1238–1247.

Hall, M.R. and Swaney, W. 2012. European Modern Earth Construction. In Hall, M., Lindsay, R. and Krayenhoff, M. eds. *Modern Earth Buildings* Cambridge UK: Woodhead Publishing: p. 654.

Jaquin, P., Augarde C. & Gerrard C. 2008. Chronological Description of the Spatial Development of Rammed Earth Techniques. *International Journal of Architectural Heritage: Conservation, Analysis, and Restoration.* Vol 2: Issue 4: pp. 377–400.

Kapfinger, Otto. 2002. *Rammed Earth.* Berlin: Birkhauser.

Lax C. 2010. *Life cycle assessment of rammed earth.* Masters thesis, University of Bath, United Kingdom.

"Lehm als Material und Medium" 2014, July 3. *Neue Zurcher Zeitung.*

Lowe K. 2012. Heaven and Earth—Sustaining Elements in Hakka Tulou. *Sustainability.* Vol 4 Issue 11: pp. 2795–2802.

Marceau, M., Nisbet, M. and VanGeem, M. 2007. *Life Cycle Inventory of Portland Cement Concrete.* Skokie, Ill: Portland Cement Association.

Morel, J., Pkla, A, and Walker, P. 2007. Compressive strength testing of compressed earth blocks. *Construction and Building Materials.* Volume 21, Issue 2: pp. 303–309.

Pérez-Lombard, L. Ortiz, J. and Pout C. 2008. A review on buildings energy consumption information. *Energy and Buildings*, Vol 40, Issue 3: pp. 394–398.

Reddy, B.V. Kumar, P. 2010. Embodied energy in cement stabilised rammed earth walls. *Energy and Buildings*, Vol 42, Issue 3: pp 380–385.

Treloar, G., Fay, O. 2001. Environmental Assessment of rammed earth construction systems. *Structural Survey*, Vol. 19 Issue 2: pp 99–106.

United States Department of Energy. 2014. Buildings Energy Databook. Compiled from *Construction Statistics of New Homes Completed/Placed.* Accessed online August 20, 2014 at: http://buildingsdatabook.eren.doe.gov.

Rammed Earth Construction – Ciancio & Beckett (Eds)
© 2015 Taylor & Francis Group, London, ISBN 978-1-138-02770-1

Who's afraid of raw earth? Experimental wall in New England and the environmental cost of stabilization

A.J. Dahmen

School of Architecture and Landscape Architecture, University of British Columbia, Canada

ABSTRACT: The use of rammed earth has grown considerably in the past half-century in the developed world, where perceptions about its environmental sustainability account for a large share of its popularity. Significant changes have accompanied its transition to the mainstream, including the use of chemical stabilizers, engineered soil blends, and mechanical placement and compaction. These alterations to the materials and installation techniques of rammed earth address the economic and structural demands of the developed world, but they also adversely effect its environmental impact. This paper assesses the effect of these changes on the embodied energy of rammed earth through a review of pertinent literature, which suggests that chemical stabilizers have the greatest effect on the embodied energy of rammed earth. The paper documents the construction of a rammed earth test wall on the campus of the Massachusetts Institute of Technology built without the use of cement stabilizers to evaluate its suitability in temperate climates. The paper offers recommendations for future research to develop a more nuanced understanding of the environmental effects of stabilized and unstabilized rammed earth in the developed world context.

1 THE COSTS OF MAINSTREAM APPEAL

1.1 Cement stabilization and embodied carbon

Rammed earth has experienced an upsurge in popularity over the past fifty years in the developed world, where it commonly incorporates cement stabilizers to provide reliable performance (McHenry, 1984; Easton, 1996; Hall & Swaney, 2012). The embodied carbon of construction materials measures the total carbon emissions resulting from raw material acquisition, manufacturing, and installation (Cohen, 2011) and is a useful measure of their environmental impact. Studies indicate that the embodied carbon of stabilized rammed earth corresponds closely with the use of cement stabilizers due to the energy required to produce cement (Treloar et al, 2001; Lax 2010, Reddy & Kumar, 2010). These findings suggest that there could be considerable environmental benefits to building with unstabilized rammed earth, referred to hereafter as "raw" rammed earth.

1.2 Changing rammed earth practices

The growing popularity of rammed earth in developed countries has resulted in changes to traditional practices. Providing predictable material performance capable of meeting the building code requirements has resulted in the addition of chemical stabilizing agents, most commonly Ordinary Portland Cement (OPC), which a study indicates is responsible for 6–7% of global carbon emissions (Chaturvedi & Ochsendorf, 2004). Where codes governing the use of rammed earth in construction have been established, such as the New Mexico (US) Building Code, they all but mandate the use of cement stabilization (New Mexico Earth Building Code, 2009).

The widespread incorporation of stabilizers is not the only change to rammed earth materials. Increasingly, utilization of site soils is replaced by imported engineered soil blends, which offer more predictable structural performance. Manual mixing and placement of soils have been replaced by diesel-powered equipment and pneumatic compaction (Easton, 1996). Generally speaking, these shifts in materials and installation methods reduce labor and increase structural reliability, but at the cost of higher embodied carbon overall (Hall and Swaney, 2012). Understanding their relative effects requires a more in depth look at the embodied carbon of rammed earth.

1.3 Embodied carbon of rammed earth

Three studies have attempted to account for the embodied carbon of rammed earth to date (Treloar & Fay, 2001), (Reddy & Kumar, 2012), and (Lax, 2010). Significantly, all three conclude that OPC stabilizers are the strongest determinant of embodied carbon of rammed earth. This finding across studies suggests that utilizing raw rammed earth offers significant reductions to embodied carbon.

An early approximation of the environmental impacts of rammed earth stabilized with 8% OPC as compared to two other forms of masonry construction in Australia by Treloar & Fay (2001) finds that OPC stabilizers are the largest contributor to embodied carbon of rammed earth. A second widely cited study on the embodied energy of stabilized rammed earth by Reddy & Kumar (2010), also finds that embodied energy of stabilized rammed earth varies linearly with OPC content. While this study is considerably more detailed than the 2001study by Treloar & Fay, Reddy & Kumar's investigation is based on manual compaction, which is rare in developed world contexts, so the findings have limited applicability to the developed world.

In contrast to the approximate methods and limited contextual relevance of the previous two studies, Lax (2010) performed a rigorous Life Cycle Assessment (LCA) study of mechanically compacted rammed earth using data from three construction sites in Great Britain. Like the first two studies, her research suggests that OPC stabilizers have the greatest effect on the embodied carbon of rammed earth, which varies directly with the amount of OPC used for stabilization. Her research suggests that at levels of stabilization of 8% (common in North America and Australia), the embodied carbon of rammed earth is two and a half times higher than that of raw earth, and within 20% that of conventional cavity wall masonry construction. At 9% stabilization, it is equivalent to that of conventional cavity wall construction. (Lax, 2010). This should be a sobering statement for those interested in rammed earth as an environmentally sustainable alternative to conventional cement-based materials.

Lax's analysis suggests that engineered soils might offer a reasonable alternative to cement-based binders. Engineered soils can address concerns about site-soil variability that can lead to unpredictable performance with a fraction of the impacts associated with OPC binders. Similarly, mechanical compaction, which according to Lax amounts to between 6–16% of the overall energy required for rammed earth, is likely an environmental tradeoff worth making, in light of the reduction to labour, increased speed of construction, and improved consolidation it produces.

2 RAW RAMMED EARTH DEMONSTRATION IN NEW ENGLAND

2.1 Motivation

A raw rammed earth demonstration wall was constructed on the campus of Massachusetts Institute of Technology in 2005 (Fig. 1) to evaluate whether rammed earth could be constructed with local soils without the use of OPC stabilizers in the climate of the Northeast United States, and to gauge the effect of the climate on the resulting construction. The wall is constructed in two sections separated by a gate. Each section measures 9 m long, 1.85 m

Figure 1. Raw rammed earth wall constructed on MIT campus (photo taken just after completion in 2005).

high, and 460 mm thick, and is covered with a steel cap that overhangs the rammed earth construction by 35 mm on either side (Figs. 1 & 2). The wall utilized engineered soil and mechanical compaction to accommodate the demands for predictable performance and to minimize labor requirements.

2.2 Soil selection and mix designs

In keeping with authorities on raw rammed earth construction (McHenry 1984, Houben & Guillaud, 1994; Easton, 1996), a well-graded mineral subsoil was sought that consisted of thirty percent non-expansive clay. Because investigations of local soils showed no easily accessible, naturally occurring soils with the necessary clay content, it was decided to engineer a soil blend that met the requirements. Non-expansive marine clays are common at a depth of 10 meters throughout the majority of the metropolitan Boston area (Terzaghi & Peck, 1996). Approximately six tons of clay was acquired from a local building site. As it was delivered in the same hydrated state in which it was excavated, a small sample was dried in an oven to determine its dry weight for use in developing mix designs. Once the mix designs were determined, the clay was mixed in its plastic state with commercially available sand and gravel. The mix designs are identified through the use of identifying steel blocks placed in the wall.

The mix designs evaluated were as follows:

– 3 parts marine clay, 7 parts structural road base (blend of stone dust and crushed stone)
– 1.5 parts marine clay, 1 part sand, 3 parts gravel (crushed granite less than 19 mm)
– 3 parts marine clay, 7 parts bank run gravel (unwashed naturally smooth stones less than 19 mm mixed with sand and fine particles.

Proctor tests established the appropriate moisture content for optimal compaction. Compressive strengths of 2 MPa (300 PSI) were produced following ASTM standards for unconfined compression.

2.3 Processing soil blends

Mixing the clay with sand and aggregates to final consistency proved to be the most challenging aspect of constructing the wall. The soil was batch mixed in a gasoline powered plaster mixer. The mixing method limited the size of aggregate used in construction; crushed stone larger than 19 mm caused problems for the machine.

In all instances the sand was washed coarse masonry sand provided by a local masonry yard. The same company delivered crushed granite. The road base came from a road repair site nearby. The marine clay was delivered in the same hydrated state in which it was excavated one to two days prior to delivery and was kept covered at all times to maintain it in a plastic state for maximum workability. It was found that the mixing the plastic clay with the various admixtures did not generally require the addition of supplementary water as the moisture in the clay provided sufficient moisture required for compaction. Appropriate moisture content of the final mix was verified in the field by the ball drop field test identified by Easton (1996).

2.4 Compaction

The first side of the wall was placed and compacted by hand hand using tampers that consisted of a a 150 × 150 mm steel plate 6 mm thick. The second half of the wall was placed and compacted mechanically with a pneumatic backfill tamper driven by a diesel compressor. Although data pertaining to the final compaction densities produced by the two methods was not collected, observation suggests that the pneumatically compacted section is denser than the manually compacted, particularly at the bottom of the lifts. Predictably, constructing the mechanically placed and compacted side went considerably faster, taking approximately one quarter of the time required for the first half.

3 DETERIORATION DUE TO WEATHER

3.1 New England climate

The wall is located in New England on the Northeastern seaboard of the United States. This region is characterized by a temperate climate that ranges between an average of 2.1 degrees Celcius in January to 26 degrees Celcius in August. Some form of precipitation falls on an average of 137 days annually. Total annual precipitation averages 1067 mm. Approximately 1118 mm of snow falls between December and March, when freezing temperatures are common. Snow often remains along the bottom third of the wall for a period of a week or more.

3.2 Deterioration due to weather

The wall has survived with no major failures since its completion during the summer of 2004. Although local areas of degradation in excess of

Figure 2. Rammed earth wall after 9 years of weathering (photo taken in 2014).

35 mm can be observed, the majority of the areas of the face have lost approximately 5–7 mm per side over the course of the nine year observation period, roughly equivalent to the the 6.4 mm of deterioration due to erosion observed by Bui et al (2008) over a period of twenty years using a stereo photogrammetric method of measurement. This would suggest that the erosion is occurring approximately twice as fast in the wall at MIT, although some of the difference could be attributable to the different methods of measurement, as well as the different climatic conditions of the two tests. One area of approximately 2 m^2 has experienced increased degradation, demonstrating a depth of material loss of approximately 35 mm. This local erosion appears to be due to large amounts of precipitation collected on the roof of the wall and running down the wall in one concentrated area. Two small areas where degradation has occurred owing to repeated exposure to water due to faulty roof are 55 mm deep over approximately 3 m^2.

The outside corners are another area of significant weathering, having lost on average approximately 30 mm of material. The relationship of different mix designs to weathering was generally inconclusive owing to variations in compaction regimens (by hand versus pneumatic) and different exposure to weather because of adjacent buildings. It was observed that higher sand contents produced smoother initial surface finishes. In some cases these sections appeared to lose less material, although this could be due to the fact that they were sheltered from the direction of the most inclement weather by adjacent buildings.

Setting aside these areas of local degradation due to increased concentrated moisture and exposure to weather, the wall might be expected to lose approximately 25–50 mm of material per side over the course of 50 years if the current trends continue. This would constitute roughly 11–22% of the total volume of the wall. However, other areas of the wall protected from direct wind-blown precipitation by adjacent buildings have experienced virtually no material loss at all. This suggests that with appropriate detailing protecting it from direct precipitation, raw earth construction could last almost indefinitely in the New England climate. This observation is supported by the significant number of extant raw rammed earth buildings in the Rhone Valley of France (CRATerre, 2006). and the Fujian province of China (Aaberg-Jørgensen, 2000), which have persisted for hundreds of years despite climates with ample of rainfall and freezing temperatures.

3.3 *Future directions*

Building with raw rammed earth in the developed world is possible and can offer significant environmental advantages over stabilized rammed earth. Capitalizing on these advantages requires further study in a number of areas outlined below.

A rigorous Life Cycle Assessment (LCA) study should be done that compares the embodied energy of raw and stabilized rammed earth between 0% OPC and 8% OPC versus other typical forms of construction. A second LCA study should be conducted to compare the total energy (embodied and operating) of a residential rammed earth structure versus the same size structure constructed with conventional materials. Such a study would put the initial embodied energy of construction into perspective with the operating energy that a study has found have found accounts for as much as 85% of total energy consumption over a thirty year period (Cole & Kernan, 1996).

Probably the most significant obstacle facing raw rammed earth construction is structural. The lower compressive strength of raw rammed earth makes it a special concern in seismic regions. One solution to this problem could be to utilize raw rammed earth with a moment frame of reinforced concrete. Comparing the environmental effects of these two approaches would require an LCA study comparing stabilized load bearing rammed earth to raw earth used as infill with supplementary structure of reinforced concrete or steel.

Finally, a comprehensive library of soils, amendments and minimum amount of stabilizers necessary to produce desired strengths should be developed to reduce uncertainty about soil performance, along the lines of the assessment criteria offered by Ciancio et al (2013).

4 CONCLUSION

Rammed earth is a minimally processed material that can offer considerable environmental benefits. However, rammed earth in the developed world is typically is stabilized with energy intensive OPC. Three studies have suggested that the embodied energy of rammed earth varies with cement content. The embodied energy of rammed earth stabilized with 8% OPC, which is common practice in the US and Australia, is comparable to conventional masonry construction.

In contrast to stabilized rammed earth, raw earth can be used to create durable buildings in the developed world without the use of chemical stabilizers. The use of engineered soils and mechanical compaction can increase reliability and reduce labor, making the use of rammed earth in the developed context possible without sacrificing the environmental advantages that often serve as its primary justification. Using raw earth presents challenges in meeting structural stability requirements, and additional research will be required realize accurate comparisons of raw rammed earth construction to stabilized rammed earth and other construction methods.

REFERENCES

Aaberg-Jørgensen, J. 2000. *Clan homes in Fujian*. Arkitekten 28: pp. 2–9.

Bui, Q. Morela, J. Reddy, B, Ghayada, W. 2009. Durability of rammed earth walls exposed for 20 years to natural weathering. Building and Environment 44 pp. 912–919.

Chaturvedi, S. and Ochsendorf, J. 2004. Global Environmental Impacts due to Cement and Steel. *Structural Engineering International* Vol. 14 Issue 3: pp. 198–200.

Ciancio, D. Jaquin, P. Walker, P. 2013. Advances on the assessment of soil suitability for rammed earth. *Construction and Building Materials* Vol. 42 pp. 40–48.

Cohen, N. 2011. Embodied Energy. In *Green Cities: an A-Z guide*. Cohen and Robbins, Ed. Sage Publications p. 165–168.

Cole, R. and Kernan, P. 1996. Life-cycle energy use in office buildings. Building and Environment Volume 31, Issue 4, July, pp. 307–317.

CRATerre 2006. *Tout Autour de la Terre*. CRATerre: Grenoble, France.

Easton, D. 1996. *The Rammed Earth House*. White River Junction, Vermont (US): Chelsea Green Publishing.

Hall, M.R. and Swaney, W. 2012. European Modern Earth Construction. In Hall, M., Lindsay, R. and Krayenhoff, M. eds. *Modern Earth Buildings* Cambridge UK: Woodhead Publishing: p. 654.

Houben, H. and Guillaud, H. 1994. *Earth Construction: a comprehensive guide*. London: Intermediate Technology Publications.

Lax C. 2010. *Life cycle assessment of rammed earth*. Masters thesis, University of Bath, United Kingdom.

Lowe K. 2012. Heaven and Earth—Sustaining Elements in Hakka Tulou. *Sustainability*. Vol. 4 Issue 11: pp. 2795–2802.

McHenry, P. 1984. *Buildings of Earth and Straw*. New York: John Wiley & Sons. p. 100.

New Mexico Housing and Construction Building Code. 2009. Title 15, Chapter 7 Part 5: Earthen Materials Code.

Reddy, B.V. Kumar, P. 2010. Embodied energy in cement stabilised rammed earth walls. *Energy and Buildings*, Vol 42, Issue 3: pp. 380–385.

Terzaghi, K. Peck, R. 1996. Soil mechanics in engineering practice Publisher: New York: Wiley.

Treloar, G., Fay, O. 2001. Environmental Assessment of rammed earth construction systems. *Structural Survey*, Vol. 19 Issue 2: pp. 99–106.

Rammed Earth Construction – Ciancio & Beckett (Eds)
© 2015 Taylor & Francis Group, London, ISBN 978-1-138-02770-1

Can we benefit from the microbes present in rammed earth?

N.K. Dhami & A. Mukherje
Department of Civil Engineering, Curtin University, Bentley, Western Australia, Australia

ABSTRACT: In recent days ability of microbes in depositing calcium carbonate in soil has been demonstrated. The technology promises to be a wonderful way to sustainable construction. This paper presents a visionary account of prospects of biological activation of soil for enhanced production of microbial carbonates and their impact on rammed earth construction. The protocols for developing a biologically activated rammed earth are discussed. The potential improvements in mechanical performance and durability are illustrated. The challenges and scope for incorporation of microbially induced calcite precipitation technology in rammed earth are mentioned.

1 INTRODUCTION

While infrastructure is a precursor to economic prosperity the present technology for building them is not sustainable. Popular materials of construction such as cement, steel and brick use too much energy and emit large quantities of Green-House Gases (GHG). The main reasons for GHG emission by these materials are high processing temperature and long distances of transportation. Moreover, these materials deplete natural resources; and they are not recycled and recyclable or reversible. Rammed earth construction that uses locally available natural materials alleviates all these problems and it is witnessing renewed interest from the engineering community due to its impeccable sustainability credentials. On the other hand, although there are excellent rammed earth constructions from the past the technique is somewhat forgotten and it faces enormous challenges to be reckoned as a main stream building technology.

Modern concrete consists of aggregates (coarse and fine) bound together by cement. It is a mixture of granular materials ranging in scale from a few 10mm to micro-millimetres in size. If one analyses the constituents of rammed earth they are not very different—gravels, stone chips (coarse aggregate), sand and clay (fine aggregates to binders) and lime (binder). The vital difference is in the binders. While cement binds the aggregates very well, its environmental cost is enormous. To produce 1Kg of cement 1Kg of CO_2 is emitted while the binders of rammed earth are produced at a much lower environmental cost. Higher strength of cement concrete does not give much dividend and rammed earth has enough strength to sustain the forces experienced by a typical human habitat. The main concern with rammed earth is, however, its higher moisture absorption,

expansion and consequent cracking (Minke, 2003). To alleviate this problem a moderate quantity of cement and occasionally asphalt emulsions have been prescribed to stabilise the traditional rammed earth (Reddy and Jagadish, 2003).

Reports from antiquity include another significant class of additives such as animal blood, molasses, eggs and animal dung. They were thought to have facilitated the workability of the mix by reducing the inter-granular friction. However, modern research shows that these additives can act as wonderful bio-activators. This aspect of the traditional rammed earth construction has remained unexplored in its modern avatars. This paper explores bio-activated rammed earth. The focus is on alleviation of the weaknesses such as moisture absorption, swelling and cracking. We shall demonstrate results from allied materials such as soil-cement blocks, sand columns and concrete blocks.

2 BIO-ACTIVATION

Microbes, especially bacteria, exist in soils at surprisingly high concentrations (around 1014 bacteria/kilogram) and their biological activities can be harnessed to improve mechanical properties of rammed earth (De Jong et al., 2013). When nutrient rich additives such as animal blood, eggs and molasses are mixed with soil the bacteria are activated. The bacterial cells act as nucleation sites and create micro niche conditions favouring precipitation of deposits within soil particles (Fig. 1). They can also reduce the inter-granular friction and improve compaction (Martirena et al., 2014). Reduced moisture absorption and increased compressive strength due to bacterial activation of sand columns have already been demonstrated

Figure 1. Carbonate crystals precipitated by bacteria along with bacterial cells.

K_{SP} is the solubility product in Eq.5.

Precipitation of calcium carbonate crystals occurs by heterogeneous nucleation on cell wall of bacteria once super-saturation is achieved and these carbonate crystals later precipitate inside the pore spaces. The bacteria not only initiate calcite precipitation, but also serve as nucleation sites for calcite crystals. Other factors such as Ca^{2+} ions, dissolved inorganic carbon, pH, and temperature in the medium influence the process.

Possible biochemical reactions in urea-$CaCl_2$ medium to precipitate $CaCO_3$ at the cell surface can be summarized as follows:

$$Ca^{2+} + Cell \longrightarrow Cell - Ca^{2+} \qquad (6)$$

$$Cl^- + HCO^{3-} + NH_3 \longrightarrow NH_4Cl + CO_3^{2-} \qquad (7)$$

$$Cell - Ca^{2+} + CO_3^{2-} \longrightarrow Cell - CaCO_3 \qquad (8)$$

(Dhami et al., 2013a). However, its application to rammed earth construction has remained unexplored hitherto. To understand the underlying phenomenon we shall briefly discuss the Microbially Induced Calcite Precipitation (MICP).

2.1 MICP

MICP is the process by which microorganisms deposit carbonates as part of their basic metabolic activities (Stocks Fischer et al., 1999). These carbonates fill the voids restricting the flow of water and gases inside and thereby reducing permeability.

Mainly four groups of microorganisms are seen to be involved in the process of MICP. They are photosynthetic organisms, sulphate reducing bacteria, organisms utilizing organic acids and nitrogen cycle. Of all the above MICP via urea hydrolysis is the simplest, most energy efficient and most widely used (De Jong et al., 2013). During microbial urease activity, 1 mol of urea is hydrolysed intra-cellularly to 1 mol of ammonia and 1 mol of carbonate (Eq.1), which spontaneously hydrolyses to form additional 1 mol of ammonia and carbonic acid (Eq.2).

$$CO(NH_2)_2 + H_2O \xrightarrow{bacteria} NH_2COOH + NH_3 \quad (1)$$

$$NH_2COOH + H_2O \longrightarrow NH_3 + H_2CO_3 \qquad (2)$$

These products equilibrate in water to form bicarbonate, 1 mol of ammonium and hydroxide ions that results in increase in pH

$$H_2CO_3 \longrightarrow 2H^+ + 2CO_3^{2-} \qquad (3)$$

$$NH_3 + H_2O \longrightarrow NH^{4-} + OH^- \qquad (4)$$

$$Ca^{2+} + CO_3^{2-} \longrightarrow CaCO_3 (K_{SP} = 3.8 \times 10^{-9}) \qquad (5)$$

MICP can be carried out in two modes, either by adding new bacterial species in the material (bio-augmentation) or by supplying nutrients to stimulate the resident bacteria (bio-stimulation). In case of rammed earth one could either mix the earth with a bacterial culture that is known for MICP (Bio-augmentation). Alternatively, one can study the resident bacterial population and stimulate them for MICP, not having to add any bacterial species (bio-stimulation). In bio-augmentation a proven bacterial strain is used. Therefore, there is a high success rate. Bio-stimulation is used to avoid addition of new species. For success, however, the resident bacterial diversity must be studied carefully and the fitness of the ureolytic strains for MICP must be ensured. Some researchers have chosen micro-augmentation where a minute quantity of new strains is used over and above the natural bacterial population.

2.2 MICP in building materials

The ability of MICP to bind granular materials is demonstrated through making sand cylinders by passing bacterial culture in sand (Fig. 2).

MICP is also proven to be effective in restoration of cement mortar cubes, repair of limestone monuments, reduction of water and chloride ion permeability in concrete, filling of pores and cracks in concrete and enhanced strength of bricks (Dhami et al., 2013c). Achal et al., (2010; 2011) investigated the effect of deposition of bacterial calcite on the surface of mortar cubes. They reported a dramatic reduction of water absorption by the mortar cube due to MICP (Fig. 3). Surprisingly, the compressive strength of the cubes also went up by about 25%. It has been observed by other investigators as well and the reason has been attributed to the bacterial mass that may have reinforced the mortar like a fibre. Understandably, diffusion of

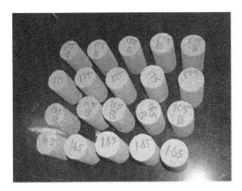

Figure 2. Sand cylinders made by MICP.

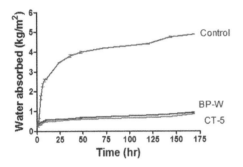

Figure 3. Reduction in water absorption due to MICP.

Table 1. Mercury intrusion porosimetry of control and bacterial treated stabilized earth blocks.

Parameter	Control block	Bacterial block
Total intrusion (mL/g)	$0.15 \pm 0.00a$	$0.09 \pm 0.00c$
Total pore area (m²/g)	$10.60 \pm 0.5a$	$6.20 \pm 0.3c$
Median pore dia (μm)	$0.01 \pm 0.00a$	$0.01 \pm 0.00a$
Av pore dia (μm 4V/A)	$0.04 \pm 0.00a$	$0.05 \pm 0.00a$
Bulk density (g/mL)	$1.82 \pm 0.06a$	$1.90 \pm 0.07a$
Apparent density (g/mL)	$2.17 \pm 0.11b$	$2.30 \pm 0.13a$
Porosity (%)	$25.30 \pm 1.7a$	$17.50 \pm 0.92b$

Experience of MICP in soil demonstrates that the technology can be extended to rammed earth. Although the authors are unaware of any recent application of MICP in rammed earth there have been some successful applications in stabilised earth blocks (Table 1).

2.4 MICP in stabilised earth blocks

Stabilised earth blocks are an excellent building material due to their low embodied energy and economy. These blocks are made by mixing soil and sand in equal proportion and compacting the material in a standard protocol. Although the blocks achieve very good structural performance they absorb moisture and consequently, they swell. High differential expansion of these blocks has led to cracking. To stabilise these blocks a moderate quantity of cement (~6%) has been used. However, cracking due to differential expansion is not avoided entirely. We made an attempt to apply MICP technology on the stabilized earth blocks by curing them with ureolytic bacterial sp. *Bacillus megaterium SS3*. 10% bacterially inoculated nutrient broth media supplemented with 2% urea and 25mM $CaCl_2$ (B-Bl) was used to make the dough. The bacterial culture with density 10^8 cells/ml was used. The block density was set at 1.8g/cc. For control blocks 10% uninoculated media was used (C-Bl) (Reddy and Gupta, 2005). Effect of MICP on water absorption, compressive strength, linear expansion and porosity has been tested. It is noticed that the saturated water content of the bacterially inoculated blocks was 40% lower than the control ones (Fig. 4a). On scanning electron microscope clear images of calcite crystals with imprint of bacteria on them was obtained (Fig. 4b). X-ray diffraction test confirmed the presence of a crystalline layer of calcite on the surface of the block (Fig. 4c). Fig. 4d presents the linear expansion in control and MICP samples. The reduction in linear expansion is seminal in alleviating cracking of the blocks.

deleterious materials such as chloride ions is also impeded by MICP. As a result, bio based sealant calcite was also effective in reducing corrosion in case of reinforced concrete (Achal et al., 2012). The bacterial deposition takes place in the pores of the substrate. Thus, the pore structure of the substrate materials is altered by MICP. The porse structure of fly ash bricks was monitored before and after bio-deposition through mercury intrusion porosimetry and it has been demonstrated that total intrusion reduced by 60% due to MICP (Dhami et al., 2012). In this case too an increase in compressive strength was observed.

2.3 MICP in soil

Bio-mediated soil improvement relies on geochemical processes that are facilitated by biological activity. Due to MICP calcium carbonate is deposited in the inter-granular spaces of soil. Thus, the grains of soil are attached together by carbonate crystal. It increases the stiffness and strength of soil. It also increases the angle of stable slope. Thus, through MICP soil can be strengthened, its slope can be stabilised and subsidence prevented. A comprehensive report of bio-geotechnical processes can be found in DeJong et al., (2013).

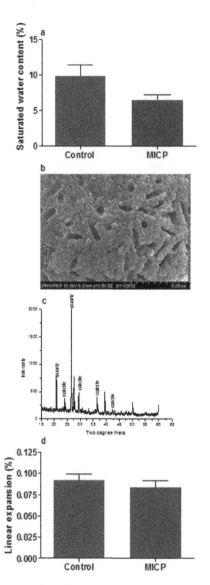

Figure 4. Effect of MICP on a) saturated water content in stabilized earth blocks b) Scanning Electron Micrograph of calcite crystals formed on the surface of stabilized earth block with bacterial impressions c) X ray diffraction pattern of crystalline layer present on the surface of stabilized earth block treated with bacterial culture d) Effect of MICP on linear expansion of stabilized earth block. Bars mean ± SD (n = 3).

3 CONCLUSIONS

This paper is an attempt to extend the experience of MICP in concrete and soil to envision its effect on rammed earth construction. Experience of MICP on concrete shows that it can dramatically reduce its moisture absorption. MICP alters the pore structure

of granular materials by depositing calcite crystals in the inter-granular spaces. Initial experiments on stabilised earth blocks demonstrate that MICP reduces the water absorption in those blocks. Thus, linear expansion due to moisture absorption and consequent cracking can be avoided. The technology looks promising for rammed earth construction. It can stabilise the rammed earth blocks, make them stronger and more durable. However, some key issues would need attention for its successful adoption. One of the main concerns would be acceptance of microbial treatment. Bio-stimulation, where no microbe is added to soil may be a preferred approach.

REFERENCES

Achal, V., Mukherjee, A. & Reddy, M.S. 2010. Biocalcification by *Sporosarcina pasteurii* using Corn steep liquor as nutrient source. *Journal of Industrial Biotechnology* 6: 170–174.

Achal, V., Mukherjee, A. & Reddy, M.S. 2011. Effect of calcifying bacteria on permeation properties of concrete structures. *Journal of Industrial Microbiology and Biotechnology* 38: 1229–1234.

Achal, V., Mukherjee, A., Goyal, S. & Reddy, M.S. 2012. Corrosion Prevention of Reinforced Concrete with Microbial Calcite Precipitation. ACI *Materials Journal* 109: 157–164.

De Jong, J.T., Soga, K., Kavazanjian, E., Burns, S., Van Paassen, L.A., Al Qabany, A., Aydilek, A., Bang, S.S., Burbank, M., Caslake, L.F., Chen, C.Y., Cheng, X., Chu, J., Ciurli, S., Esnault-Filet, A., Fauriel, S., Hamdan, N., Hata, T., Inagaki, Y., Jefferis, S., Kuo, M., Laloui, L., Larrahondo, J., Manning, D.A.C., Martinez, B., Montoya, B.M., Nelson, D.C., Palomino, A., Renforth, P., Santamarina, J.C., Seagren, E.A., Tanyu, B., Tsesarsky, M. & Weaver, T. 2013. Biogeochemical processes and geotechnical applications: progress, opportunities and challenges. *Geotechnique* 63: 287–301.

Dhami, N.K., Mukherjee, A. & Reddy, M.S. 2012. Improvement in strength properties of ash bricks by bacterial calcite. *Ecological Engineering* 39: 31–35.

Dhami, N.K., Mukherjee, A. & Reddy, M.S. 2013a. *Bacillus megaterium* mediated mineralization of calcium carbonate as biogenic surface treatment of Green building materials. *World Journal of Microbiology and Biotechnology* 29: 2397–2406.

Dhami, N.K., Mukherjee, A. & Reddy, M.S. 2013b. Biomineralization of Calcium Carbonate Polymorphs by the Bacterial Strains Isolated from Calcareous Sites. *Journal of Microbiology and Biotechnology* 23: 707–714.

Dhami, N.K., Mukherjee, A. & Reddy, M.S. 2013c. Biomineralization of calcium carbonates and their engineered applications: a review. Frontiers of Microbiology 4: 314.

Martirena, F., Rodriguez-Rodriguez, Y., Callico, A., Gonzalez, R., Diaz, Y., Bracho, G., Alujas, A., Guerra de Leon, J.O. & Alvarado-Capó, Y. 2014. Microorganism—based bioplasticizer for cementitious materials. *Construction and Building Materials* 60: 91–97.

Minke, G. 2003. Earth construction handbook, WIT press, Boston.

Rammed Earth Construction – Ciancio & Beckett (Eds)
© 2015 Taylor & Francis Group, London, ISBN 978-1-138-02770-1

Preparing regulatory challenges and opportunities for small to medium residential scale stabilized rammed earth buildings in Canada

K.J. Dick
University of Manitoba, Winnipeg, Manitoba, Canada

T.J. Krahn
Building Alternatives Inc., Codrington, Ontario, Canada

ABSTRACT: Since 2005, the National Building Code of Canada has contained an alternative solution application protocol, allowing novel materials and designs to be proposed to meet the objective and functional code statements that express the full intent of the legislation. Each province and territory adopts the model national code with various additions, omissions and amendments as they see fit. The code is enforced at a municipal level, and is frequently accompanied by region specific by-laws. Larger, more densely populated municipalities generally have larger, more sophisticated plan examination departments, often employing professional engineers and architects to vet permit applications that fall outside of prescriptive solutions. This leads to apparent discrepancies in the logistics of plan examination, permit administration, inspection and final occupancy stages of construction between larger and smaller municipalities. The challenge for builders and designers is to put together permit submittal packages that satisfy professional engineers and architects who are not familiar with the materials or methods of construction being proposed. Demonstration of equivalence to an existing design standard for a similar material is frequently required by code officials—in the case of rammed earth, conformance to concrete or masonry design methods. Materials testing to confirm assumed design strength before permit granting is required, along with testing during construction to assure quality in the actual building. The challenge for regulatory bodies is to provide fair and timely review of submissions and to guide proponents along a code and by-law compliant path without actively joining the design team. All of these challenges can also be expressed as opportunities for collaboration between design professionals, builders and regulatory officials. This paper discusses these challenges using actual practice examples where appropriate. The authors also propose recommendations with regards to the design-approval process.

1   INTRODUCTION

Authorities having jurisdiction in Canada are currently in their second code cycle since the introduction of an objective-based national model code. The first National Building Code of Canada (NBCC) to adopt an objective-based format was issued in 2005. The Canadian Commission on Building and Fire Codes attempts to re-issue an updated version of the major codes (Building, Fire, Plumbing & Electrical) every 5 years. The current national model building code is the 2010 edition, with a 2015 edition on pace to be issued late in 2015 or early 2016. (Canadian Commission on Building and Fire Codes, in press)

The move to an objective-based code did not eliminate the listing of prescriptive solutions for a given building assembly, rather it involved adding alternative regulatory paths to acceptable solutions. By defining the goals of the code via cross-referenced objective and functional statements, the objective-based format attempts to give designers and code officials methods to evaluate a potential design for conformance apart from a 'cook-book' approach. Specifically, an alternative solutions proposal protocol was introduced into the 2005 NBCC. However, differences in the way that each province and territory adopts the model code into their legislation, compounded with differences in the way any given municipality enforces their regional code and/or modifies it via local by-laws, leave designers and project proponents with a range of conditions to deal with when applying for a permit to build.

Before the adoption of the objective-based model code, non-conforming materials and designs were permitted on a project by project basis, either via the building official's discretion, via some type of approved research program, or because of exceptional circumstances. An example of the building official's discretion is given in the first case study below. An approved research program is most often a case where a municipality and an academic institution cooperate to demonstrate a novel building technique that is funded publicly. Exceptional circumstances are really an extreme case of this; for instance, an Olympic village or World's Fair site. It is not the purpose of this paper to deal with projects of that magnitude per se, rather the example is given because those projects are also designed, permitted, insured and funded—simply at a scale much higher than small to medium scale residential builds.

Both before and after the advent of the objective-based code model, a key element to winning the building official's approval directly or via a research program is the establishment of material qualities that can be measured and shown to be consistent with the design methodology adopted by the engineer or architect. This is a primary challenge for the designer; choosing an accepted design methodology developed for a similar, yet different, material and then developing a test method to prove that the different material behaves sufficiently like the accepted one to justify the analysis and final design. Two examples are given in the case studies, one following the Canadian concrete design manual and the other the Canadian masonry design manual. Engineering design standards and their accompanying manuals, guides and commentaries are published by the Canadian Standards Association (CSA).

The Canadian Construction Materials Centre (CCMC) is responsible for evaluation and national certification of innovative building materials, products and systems. Conventionally, a material or product attains CCMC certification in order to be widely accepted by designers, regulators and builders. All CCMC certifications are referenced in the NBCC by default, allowing relatively easy specification and acceptance. At some point in the future, the material or product may be cited directly in the body of the building code itself. Polystyrene Insulated Concrete Forms (ICFs) are an example of a product/system that has gone from CCMC evaluation to outright specification in the national code within the past 20 years.

The variability in aggregate content, mix recipe and method inherent in working with natural or pre-industrial materials and techniques such as SRE effectively precludes evaluation by a body like the CCMC. It should also be noted that the evaluation process is lengthy and expensive, and to date there have not been any proponents of earthen construction in Canada willing to attempt it.

This leads to each project being evaluated on its own merits, and raises another challenge. Before the issuing of a permit, inspection criteria must be determined along with a quality control and materials testing program to be carried out during construction. A requirement for any project varying from common construction techniques or materials is a Commitment to General Reviews by the design professionals. This is a basic form establishing the party or parties responsible for inspections and site reviews, but it does not include a great amount of detail and is often accompanied by a document clearly stating the agreed upon schedule, notable milestones and substantial completion criteria. Two different pre-construction testing programs and two different construction phase inspection protocols are given in the case studies.

2 CASE STUDIES

2.1 Extra-urban residence—Huntsville, Ontario

The Allen residence, located just outside of the town of Huntsville, in the Muskoka region of Ontario, was completed in the fall of 2012. It was the first Stabilized Rammed Earth (SRE) single-family dwelling to apply for a building permit in the region. The pre-construction materials testing program for this project was initiated in the spring of 2010.

The town of Huntsville has a Development Services branch, which includes their Building and Planning Departments, along with By-Law enforcement and Sustainability. As of 2013, the Building Department did not employ any registered professional engineers for plan examination or inspection.

The design methodology for the engineering of the SRE walls on the Allen residence was a hybrid of the Canadian Concrete Design standard CSA A23.3 (Canadian Standards Association, in press) and various techniques and analysis tools taken from the international literature. Of primary concern was the effect of freeze-thaw cycles on exposed SRE walls in a Canadian climate. Pre-construction testing included evaluating different grain size distributions in the source soil mix, varying Portland cement content, the addition of a silicon emulsion ad-mixture for permeability reduction (Plasti-cure by Tech Dry), and oxides for colour control. The structural design was controlled in large part by the compressive strength of the test samples. 150 mm diameter × 300 mm tall test cylinders were tested at 28 and 56 day curing times. The durability of the different mixes was tested by exposing block style samples to the environment and by creating excessive freeze-thaw cycles during the winter months.

For durability, a minimum Portland cement content of 5% by weight, plus the manufacturer's recommended dosage of admixture to reduce permeability was determined to be adequate. For structural stability, a minimum of 7.5% Portland cement by weight was determined to be necessary to achieve a 15 MPa design compressive strength. The testing program was summarized in a simple document and presented to the building officials in Huntsville along with the completed drawings set at the time of permit application. The building official requested an in-person meeting with the structural engineer in order to discuss the material and building technique, and was satisfied within 15 minutes; provided that the engineer take on responsibility for inspecting the SRE walls and assume full liability for their performance.

Follow-up testing was requested, involving samples taken during construction but occupancy was not denied before the test results were submitted after construction was completed.

2.2 Urban residence—Ottawa, Ontario

The Smyth-Allcott residence is a two storey single family dwelling currently under construction just south of Ottawa, Ontario. The building has single storey stabilized rammed earth walls with light wood frame second storey walls above. The conceptual design was taken to the city of Ottawa's building department for an initial consultation by the client and architect in April of 2012. At that point in time, a zoning official looked over the proposed design and did not see any outstanding issues that would prevent or delay a building permit being issued.

Following a similar pre-construction testing program to the one employed for the Allen residence, the design was completed over the winter and spring of 2012/13. The city of Ottawa is in a seismic zone, and the appropriate lateral load capacity of the structural walls is required to be shown in any engineering design submitted for permit.

The city of Ottawa has a Planning and Growth Management Department, employing several registered professional engineers and architects in the Building Code Services division. In the case of the Smyth-Allcott permit application, a technician in the residential plan examination division reviewed the plans and then passed them up to an engineer in the commercial division. The technician did not feel qualified to review the plans, as the structure included materials and techniques outside of part 9, the prescriptive core of the Ontario Building Code (OBC).

The structural engineer reviewing the set of plans and calculations was not familiar with earthen construction methods, nor with stabilized rammed earth as a material that could be designed using engineering principles. As a result, the engineer requested evidence via testing done in a Canadian context to prove that stabilized rammed earth could

reliably be designed in general accordance with CSA A23.3. Notwithstanding the lack of published research on stabilized rammed earth from Canada, two larger concerns were raised regarding the use of the concrete design standard for this different material. First, the minimum compressive strength for reinforced concrete is currently set at 25 MPa. Second, the CSA A23.1 and A23.2 standards (Canadian Standards Association, in press) set limits to the quantity of particles of less than 80 μm diameter present in a given sample of aggregate. The inability of SRE to meet these two qualities effectively ruled out the use of the concrete design standard for engineering analysis in this case.

Supporting documents submitted with the initial permit application included the New Zealand Engineering Design of Earth Buildings (NZS 4297:1998), which is written in concert with both masonry and concrete design methodologies for reinforced and un-reinforced assemblies alike. Reference to this standard prompted a re-design carried out in general accordance with the CSA S304.1 standard, Design of Masonry Structures (Canadian Standards Association, in press). The minimum compressive strength for reinforced masonry under seismic loading in CSA S304.1 is 15 MPa, and the standard contains no minimum aggregate size criteria, as masonry containing clay—both fired and chemically stabilized—are permitted.

The primary change to the engineering analysis resulting from the shift to a masonry-based standard from a concrete one was the increased importance in slenderness ratio as opposed to reinforcement in driving the final design.

In terms of the permit application process, the fastest path forward was determined to be an alternative solution application asserting the equivalence of the reinforced stabilized rammed earth wall to a reinforced masonry wall.

3 CONCLUSIONS & RECOMMENDATIONS

It is common to encounter different interpretations of building code requirements between different jurisdictions. It is also common to find that a jurisdiction with a larger population will have more rigorous plan examination and inspection requirements than a neighbouring jurisdiction with a smaller population. In large part, this is due to higher staffing capacity and experience with a broader variety of projects in the more populous region. However, the resulting inconsistency in the application and enforcement of federal and provincial codes at the municipal level effectively creates separate classes of construction regulation where no such separation is intended, or even allowed.

Our recommendation is for regulators to set a minimum level of adjudication necessary for an alternative solution proposal to be considered. This may involve third party professionals in some jurisdictions, but this is not unprecedented for plan examinations or inspections that are outside the expertise of the staff in any given building department.

Correspondingly, designers must educate themselves about what regulators need to see in order to move a permit forward when an unfamiliar material or building technique is being proposed.

The challenge lies in working together without blurring lines of liability and client responsibility. At the same time, the opportunity exists to work together to create clear and consistent design and administrative guides that lead to a better built environment.

REFERENCES

Canadian Commission on Building and Fire Codes. *National Building Code of Canada, 2005 edition* (National Research Council of Canada, Ottawa, 2005).

Canadian Commission on Building and Fire Codes. *National Building Code of Canada, 2010 edition* (National Research Council of Canada, Ottawa, 2010).

Canadian Standards Association. CAN/CSA A23.1-09— *Concrete materials and methods of concrete construction* (Canadian Standards Association, Mississauga, July 2009).

Canadian Standards Association. CAN/CSA A23.2-09— *Test methods and standard practices for concrete* (Canadian Standards Association, Mississauga, July 2009).

Canadian Standards Association. CAN/CSA A23.3-04— *Design of concrete structures* (Canadian Standards Association, Mississauga, July 2007).

Canadian Standards Association. CAN/CSA S304.1-04— *Design of Masonry structures* (Canadian Standards Association, Mississauga, December 2004).

OBC (2006) *Ontario Building Code*, (Ministry of Municipal Affairs and Housing, Markham, 2006).

OBC (2012) *Ontario Building Code*, (Ministry of Municipal Affairs and Housing, Markham, 2012).

Rammed Earth Construction – Ciancio & Beckett (Eds)
© 2015 Taylor & Francis Group, London, ISBN 978-1-138-02770-1

Techniques of intervention in monumental rammed earth buildings in Spain in the last decade (2004–2013). 1% Cultural Programme

L. García-Soriano, C. Mileto & F. Vegas López-Manzanares
Instituto de Restauración del Patrimonio, Universitat Politècnica de València, València, Spain

ABSTRACT: This research work proposes an analysis of the techniques proposed in the interventions carried out in the rammed earth architecture in Spain in the last decade according to the of 1% Cultural Program. Since the analysis focuses on interventions funded by public funds, the selected study sample consists exclusively of monumental buildings, and there are outside the scope of this paper the rammed earth buildings of vernacular architecture. The research presented here consisted in reviewing the collection of the archives of the Ministry of Development, with a view to providing an initial approach to the interventions on buildings made of rammed earth in the last ten years (from 2004 until 2013) and funded by Spanish Government. The methodology for analysis is the documentation with technical sheets to study the characteristics of each intervention. These records allow us to perform an analysis of similarities and contrasts in interventions which have a same building technique.

1 INTRODUCTION

1.1 *Aim of the research*

The main aim of this research is to provide an overall analysis of the interventions carried out on monumental rammed earth architecture in Spain through the Spanish government's 1% Cultural Programme over the last decade.

The case studies examined are monumental buildings given that these interventions receive public state funding, and rammed earth constructions of vernacular architecture are excluded from the analysis.

The main goal of this study was to focus the analysis of these interventions mostly on the construction techniques proposed for the actions and for a specific area of wall, the union between the original material and the new material used for intervention. The analysis of the different suggestions for the resolution of this bond demonstrates the similarities and differences in contemporary interventions.

1.2 *The 1% Cultural Programme*

The 1% Cultural Programme was included in the 1985 Historical Heritage Law, which stipulated that at least 1% of the budget of public works contracts was to be devoted to work towards conserving and enriching Spanish historical heritage.

Most of the funds in this programme, developed in collaboration with the Ministry of Culture, come from the Ministry of Development (Sánchez Llorente 2010). A joint committee of the Ministries

of Culture and Development was established and an interministerial agreement was set up defining criteria for action and priorities in interventions to be carried out. In recent years, this programme has mainly been applied to the execution of conservation and restoration work of Spanish historical heritage buildings, and its budget is even higher than that provided by the Secretary of State for Culture (Lafuente Batanero 2004).

The main applicants for this subsidy are local administrations (town councils) themselves, although applications are also made by the provincial or regional administrations and foundations and religious congregations.

1.3 *Research methodology*

For this research on the interventions funded by the 1% Cultural Programme in rammed earth architecture in the last decade we have worked with material from the archives of the Ministry of Development, where the intervention projects funded by this programme from the 2004 mixed committee until the present can be found.

The general list of all the interventions carried out in this period (627 in total) was used as a starting point, and a selection was made of the interventions on buildings originally executed using the rammed earth technique. 75 files were selected. It should be noted that as direct action on the walls is not carried out in all intervention projects the definitive samples used for this research consist of 68 buildings distributed fairly homogeneously throughout the Spanish territory in the areas where monumental rammed earth architecture is commonly found (Fig. 1).

It is also important to note that most of the case studies are buildings classified as military architecture (castles, defensive walls, towers ...) but there is a small group which includes buildings with civic and religious architecture (palaces, churches, convents ...).

A table was drawn up for the analysis showing the actions executed for each case study. To structure the table distinctions were made between superficial interventions and deep interventions (those which affect most of the thickness of the wall).

Another aspect taken into account was that of prior cleaning actions in the wall to be repaired, that is to say, whether these are simply cleaning actions or whether they involve the elimination of material in order to achieve a regular surface. Moreover, and this may well be the most important aspect, the bonding elements used to improve the anchoring of the new material to the original material have been analysed.

Overall conclusions can be drawn from a cross-analysis of the different interventions provided by the table for each case study (Fig. 2).

Figure 1.   Geographic distribution of the study cases in autonomous communities (L. García Soriano).

2   ANALYSIS OF THE INTERVENTIONS

Following the collection of data and the creation of the table an overall analysis of the techniques used in the different case studies was carried out.

In all cases, the original rammed earth construction technique is recommended for the restoration or reconstruction of the wall using materials that are similar to the originals, both in the fabric and formwork (*tapial*). This intended use of materials similar to the original ones also extends to the use of formwork and follows the metrics of the existing construction. Accordingly, in the 2007 intervention in the Castle of Anento (Zaragoza) it was proposed that "in the areas in which the replacement of volume is small, and on the lower section, the extrados of the missing parts should be constructed using lime rammed earth. Measurements should be taken on site of the imprints of the formwork of the existing rammed earth, and new wooden formwork with the same measurements should be used to achieve a similar overall final texture" (file 02-50028-01940-10).

Nevertheless the use of white cement in the mix is proposed for many interventions in order to enhance the features of the earth mix, as in the case of the Castle of San Juan in Calasparra (Murcia) where a proposal is made for the "reconstruction of the existing rammed earth face (15 cm average thickness) with materials similar to the original ones, using a base paste prepared with natural aggregates, stabilised with slaked lime and a minimum proportion of white cement and colouring if necessary" (file 13-30013-01908-09) (Fig. 3).

It can therefore be stated that in these intervention projects on rammed earth buildings the construction technique used is essentially the original one and the materials proposed are similar to those existing both in the filling of gaps (superficial or deep) and in the partial reconstructions executed in some cases, generally of structures built on the crowning of the wall. Although the technique and materials are similar to the original ones, new materials that are very different to the original ones should be introduced into the bonding elements.

Figure 3.   Castle of San Juan in Calasparra (Murcia) (L. García-Soriano).

	Surface interventions			Major interventions (volume restitution)		
	Disintegrated material removal	New material similar to the original	Connector element	New material similar to the original	Connector element
Reina Castle	x	x	Steel mesh	x	Galvanised steel rods and mesh
Moratalla Castle		x	----	x	Stainless steel rods
Castle of Miraflores (El Burgo)	x	x	Glass-fibre mesh and wooden needles every 50 cm	x	Glass-fibre mesh and wooden needles every 50 cm
Alhama de Murcia Castle	x	x	Glass-fibre mesh	x	Glass-fibre mesh

Figure 2.   Example of some of the cases analysed in the table (L. García-Soriano).

2.1 Analysis of the bonding elements between new and original materials

This research focuses mainly on a specific part of the wall, the bonding between the new and original material. This analysis will help decipher the solutions proposed in the last decade for the projects executed and will provide an answer to specific problems, which if incorrectly resolved, could lead to many later pathologies in the wall (detachment, corrosion, material incompatibility ...).

Two distinct types of action can be identified: actions which propose a physical union (mortise joints in a wall allowing a tongue and groove union between the new material and the original one) and interventions proposing a union using auxiliary anchoring elements (bars, mesh, etc.) in some cases where there was already mortise.

In most cases, when the project proposes the reintegration of elements missing from the wall that are essential to its cross-section, the option chosen is the use of bonding materials that facilitate the anchoring of the new material to the original material. It should be noted that these are complex actions to be executed with formwork solely on one face.

In a small group of interventions (approximately 20%) steel is chosen to resolve this union, as in the case of the 2005 intervention on the Castle of Nogalte, where it is stated that this is a "consolidation of the existing wall (...) including internal reinforcement using a vertical grid of 3 mm diameter threaded steel rods" (file 13-30033-00562-04) (Fig. 4). Another example of the use of steel is the 2006 intervention in the Alcazaba de Reina which proposed that "to improve adhesion to the original fabric a galvanised steel structure should be incorporated to anchor wherever necessary" (file 10-06110-01440-06) (Fig. 5).

Material compatibility between steel structures and the earth used in rammed earth walls is not clearly determined. This may be the reason that fibreglass is generally chosen to execute these unions. For instance, in the case of the 2001 intervention in the Castle of San Juan in Calasparra,

a proposal was made to reconstruct the rammed earth wall "executed on the existing constructions, previously cleaned and treated, and anchoring the new construction to the original one using fibreglass rods (10 mm diameter every 40 cm) and a fibreglass mesh in 6 mm grids" (file 13-30013-01908-09).

Another important aspect to take into account is that of the tasks for the cleaning and treatment of the wall prior to the execution of new walls. For all cases it is advisable to clean and consolidate the damaged surface on which the reintegration is to be performed. Usually once it has been cleaned the new wall is executed respecting the original deteriorated profile, but in some cases the decision is made to eliminate material to form mortise joints of regular surfaces to which new material may be anchored. An example of this is the 2009 intervention in the Castle of Miraflores in El Burgo in which mortise joints in the wall and a solution combining fibreglass anchoring with wooden stakes are proposed, stating that "... after cleaning the section to be restored as far as the undamaged rammed earth, cross-bars of dry branches of wood are staggered every 50 cm to support a wide fibreglass mesh which after formwork will be filled with 10 cm layers ..." (file 01-29031-01878-09) (Fig. 6).

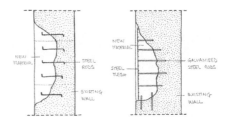

Figure 5. Details of some of the solutions proposed for the union of new materials with original materials. Left: Detail created using the documentation from file 13-30028-01803-09. Right: Detail created using the documentation from file 10-06110-01440-06 (L. García-Soriano).

Figure 4. Castle of Nogalte (Puerto Lumbreras) (L. García-Soriano).

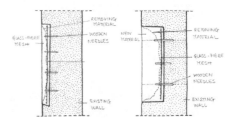

Figure 6. Details of some solutions proposed for the bonding of the new material to the original material in different thicknesses. Detail created using the documentation of file 01-29031-01878-09 (L. García-Soriano).

When the reintegrations are more superficial with only a few centimetres of missing material, the method proposed is generally similar to that of deep reintegrations, using the same bonding element, with the sole exception that in these cases the reintegration is not executed with formwork and rammed earth, but with 1 or 2 cm thick successive layers of mortar (usually earth and lime) applied to the bonding element.

If we analyse the union of the new walls in the reconstructions of crownings, these are generally built on top of the existing ones. Although in some cases like that of the 2009 intervention in the Castle of Cervera del Maestre construction unions are proposed which use mortise joints for improvement, with the project stating that "in many areas of the rammed earth wall parts of the outer shell and nucleus have become detached, up to a depth ranging from 30 cm to one metre. This part of the wall is to be restored using stonework and a concrete mix of white cement and rich lime similar to that in place. In the vertical joins of the rammed earth walls the adhesion is improved by emptying the central area of the union and later filling it in with the materials and stonework from the next rammed earth wall" (file 17-13044-01718-19).

3 CONCLUSIONS

Following this analysis a series of preliminary conclusions are reached regarding the interventions funded by the 1% Cultural Programme in the last decade.

In the first decade of the 21st century many interventions, commissioned by the 1% Cultural Programme and fairly homogeneously distributed throughout the Iberian Peninsula, were carried out on rammed earth buildings.

As regards the construction technique and details of the interventions carried out it is interesting to note that the traditional rammed earth technique is mainly used in intervention, employing the construction variation of the original walls. However, in most cases the unions between the original material and the interventions are proposed with union elements using modern materials. It can therefore be stated that in these interventions the general trend is towards using traditional construction techniques (rammed earth) while incorporating new materials at specific points (and at times also in the mortar). This is perhaps because new materials are better known, easily accessible and also considered to have superior technical characteristics which will improve the adhesion of the new rammed earths. The effect of the material compatibility or incompatibility of these elements and their evolution over time is as yet unknown.

This study aims to be an initial approximation and preliminary analysis providing general information on a group of interventions and on specific aspects of these, the solutions proposed to join the material used in the intervention with the original rammed earth walls. This is why it should be understood as a first step within more extensive ongoing research which analyses other aspects relating to construction techniques, intervention criteria, pathologies, etc.

ACKNOWLEDGEMENTS

In order to carry out this research, we are grateful for the help of Rita Lorite, coordinator of the 1% Cultural Programme, and Laura Collado, Head of the Documentation and Archive Department, which has been fundamental in giving us access to the archives of the Ministry of Development concerning the works funded under this programme, at the Sub-Department General of Architecture and Building.

This study is part of a research project funded by the Spanish Ministry of Science and Innovation "La restauración de la arquitectura de tapia en la Península Ibérica. Criterios, técnicas, resultados y perspectivas" (Ref.: BIA 2010-18921; main researcher: Camilla Mileto) and is part of the research work of "La restauración de la arquitectura de tapia de 1980 a la actualidad a través de los fondos del ministerio de Cultura y del Ministerio de Fomento del Gobierno de España" by doctoral student L. García-Soriano.

REFERENCES

Dossier 01-29031-01878-09 of the Ministry of Development Archives. *Intervención en el Castle of Miraflores (El Burgo—Málaga)*.

Dossier 02-50028-01940-10 of the Ministry of Development Archives. *Consolidación y Restauración de los restos del Castle of Anento (Zaragoza)*.

Dossier 10-06110-01440-06 of the Ministry of Development Archives. *Rehabilitación de la Alcazaba Arabe, Fase I: Consolidación del Frente Noroeste. (Reina—Badajoz)*.

Dossier 13-30013-01908-09 of the Ministry of Development Archives. *Rehabilitación y Consolidación del Castle of San Juan (Calasparra)*.

Dossier 13-30033-00562-04 of the Ministry of Development Archives. *Rehabilitación del Castle of Nogalte o de Puerto Lumbreras*.

Dossier 17-13044-01718-19 of the Ministry of Development rchives. *Consolidación de la torre y muralla norte del Castle of Cervera del Maestre, Fase III*.

Lafuente Batanero, L., 2004. Las medidas de fomento. Aplicación de la nueva Ley de Mecenazgo en los museos [*Página web del Ministerio de Cultura: Subdirección General de Museos Estatales*], [Online].

Sánchez Llorente, A., 2010. El 1% Cultural. Una visión práctica. *Revista de Patrimonio Cultural de España* 3 (La economía del patrimonio cultural), pp. 129–142.

Rammed Earth Construction – Ciancio & Beckett (Eds)
© 2015 Taylor & Francis Group, London, ISBN 978-1-138-02770-1

Investigation of energy performance of a rammed earth built commercial office building in three different climate zones of Australia

M.M. Hasan
Queensland University of Technology (QUT), Brisbane, Queensland, Australia
Member of Engineers Australia, Member of ABSA, Australia

K. Dutta
Curtin University, Perth, WA, Australia

ABSTRACT: This paper will examine the predicted Annual Energy Consumption (kWh/yr) of a commercial office building made of rammed earth in Sub-tropical, tropical and temperate climate of Australia using design data of the building fabric and glazing. An in-depth analysis for the energy efficiency of the building is cost effective and considered as a best practice for building design and construction prior to the commencement of the project. To improve Energy performance of this building project, thermal simulation and comprehensive simulated data analysis were conducted, using DesignBuilder. This includes detailed analysis of Annual Energy Consumption (kWh/yr), including Heating and Cooling Energy (kWh/yr). Different constructions, including rammed earth wall, lightweight wall and heavy weight wall were used for the evaluation of the proposed office building's thermal simulation in order to achieve a cost effective and energy efficient solution. Finally, changes to any building elements for Energy performance improvement of the building project are identified, and its compliance with National Construction Code (NCC) of Australia for three climate zones is assessed.

1 INTRODUCTION

Buildings worldwide account for a surprisingly high 40% of global energy consumption (Energy Efficiency in Buildings, 2009). Both residential and commercial buildings account for approximately 23% of Australia's greenhouse gas emissions (Building and Construction, 2011). Heating, Ventilation and Air-Conditioning (HVAC) consumes nearly 33% of total energy consumption of commercial buildings in Australia (CIE, 2007). Building fabric, particularly, type of building constructions and glazing play a key role to reduce the energy consumption of the building. Ciancio and Beckett (2013) highlighted the use of sustainable building material such as Rammed earth wall to reduce the use of HVAC and to achieve a comfortable living space. Rammed earth walls have low thermal resistance, but high thermal mass compared to light weight construction. However, thermal resistance is not the only factor responsible for providing a comfortable living environment (Allinson & Hall, 2007; Faure & Le Roux, 2012). Studies in New South Wales, Australia and in West Argentian, Galcia and Spain (Page et al. 2011; Larsen et al. 2002; Orosa & Oliveira 2012) indicated that

the high thermal mass but low thermal resistance provides better thermal performance and therefore lower heating and cooling demand when compared to high thermal resistance materials. In Australia, a hypothetical un-insulated rammed earth built house was investigated by using AccuRate software in climate zone 3, 5 and 7 (Dong et al. 2014). Energy consumption of a rammed earth built office building was studied in Charles Sturt University in New South Wales, Australia using questionnaire survey and simulation (Taylor et al. 2008). However, a comprehensive study on Annual Energy Consumption scenario and its compliance with building code is required for rammed earth built commercial office building for different climate zones of Australia. In this study, Energy performance in terms of heating and cooling of a rammed earth (R value 0.32 m^2.K/W and thermal mass 1285 KJ/m^3K) built commercial office building is examined using design data of the building fabric and DesignBuilder simulation before the building construction. As per Section J of NCC 2014, new commercial building must need to satisfy the Section J Performance requirement (JP1) that includes the building fabric during the design stage of the building. To serve this purpose, design compliance for energy efficiency of a single storey commercial office building (17 m × 8 m) is being assessed by Section J Verification method (JV3) using DesignBuilder thermal simulation with different construction details including lightweight, heavy weight and rammed earth construction. The Energy performance of the rammed earth built commercial office building is also compared with light weight and heavy weight constructions in Sub-tropical (Climate zone 2), tropical (Climate zone 1) and temperate climate (Climate zone 6) of Australia.

2 METHODOLOGY

DesignBuilder version 3.4 software that satisfies Australian Building Code Board (ABCB) protocol and uses EnergyPlus version 8.1 engine, has been used in this study to investigate the energy performance of a commercial office building. First, all architectural design data including floor plan, elevations, sections, site plan, wall and roof constructions, glazing (all external glass doors and windows) and finishes schedules (Light color for wall and roof, Solar absorptance (α) = 0.4) have been collected. The typical floor plan is shown in Figure 1.

For the office building in Sub-tropical climate, a location has been fixed such as Brisbane (-27°S 153°E). Using construction details, 3D modeling (Fig. 2) and zoning of the designed building in DesignBuilder software have been completed. 10 different types of wall constructions as shown in Table 1 have been used in this study.

10 separate simulations have been conducted to obtain the heating, cooling and total annual energy consumption of the building for these 10 types of constructions. Then the energy consumption of the building for 10 different constructions has been compared with Reference buildings (NCC compliant) to make the proposed building energy efficient. After that, 10 Reference buildings for these 10 dif-

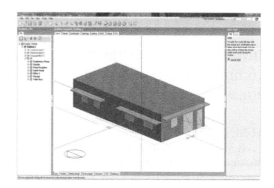

Figure 2. 3D Model of the Building in DesignBuilder.

Table 1. Wall constructions used in DesignBuilder simulation.

Number in results (No)	Type of construction
1	Cavity Panel
2	Cavity Panel with R1.5 insulation
3	190 mm concrete slab/ block
4	190 mm Concrete panel + Plasterboard
5	150 mm slab
6	Stabilised Rammed Earth (SRE)
7	Stabilised Rammed Earth + Plasterboard
8	Insulated SRE R2.5 total (ISRE)
9	Brick Veneer wall
10	Brick Veneer wall with R1.5 Insulation

ferent constructions are modeled in DesignBuilder following JV3 assessment criteria and conditions to comply with section J Energy Efficiency of NCC. The above procedure is followed for a location (Cairns) in tropical climate and a location (Melbourne) in temperate climate. Finally, a comparison is made between rammed earth built commercial building and other types of construction based on their energy efficiency. A change of any building elements such as insulation requirement or glazing are analyzed and suggested to comply with NCC Section J Energy Efficiency.

3 RESULTS

A sample result in DesignBuilder thermal simulation for Annual Energy Consumption (AEC) in terms of Electricity of Stabilised Rammed Earth,

Figure 1. Floor plan of the proposed office building.

SRE (No 6) in Subtropical climate is shown in Table 3. The numbers represent the cumulative energy transfer totals relative to conditioned spaces. Positive numbers indicate energy transfers from outside to inside of the building and negative means the vice versa for a building component.

Room electricity and lighting are kept same for all thermal simulations. The building component that affects the thermal simulation results in this study for the proposed commercial office building is walls. Change of wall constructions affect the energy transfer number for other building fabrics such roof, floor, glazing, and solar heat gain coefficient of exterior windows. The Reference buildings differ from proposed buildings in terms of insulation, glazing and colour of wall and roof (Table 2).

3.1 Sub-tropical climate

In Sub-tropical climate, 9 out of 10 wall constructions of the proposed building have lower energy consumption than the DTS-compliant Reference Buildings (Fig.3). The results indicate that these constructions are energy efficient, and satisfy the criteria of Section J of JV3 assessment of the NCC 2014 at the same time. Rammed earth built constructions (No 6, 7) of this commercial office buidling performed better than other light-weight constructions (No 1, 2, 9 and 10) as shown in

Table 2.   JV3 assessment criteria and reference building conditions.

JV3 Assessment criteria

a. 3D model of the building with location and orientation
b. Schedules for: occupancy, internal heat loads, lighting and HVAC system Simulation hours: 8760 hours, at least 2500 hours/year.
c. Thermostats setting: 18°C to 26°C.
d. Air conditioning and Artificial Lighting complies: NCC Parts J5 and J6.
e. The air conditioning & heating Annual Energy Efficiency Ratio (AEER): NCC Performance Requirement JP3, Cooling AEER: Minimum Energy Performance Standards (MEPS). HVAC Design Factors: 1.0 for 98% coverage
f. The fresh air rate: 10 L/sec/person. Infiltration: 1.0 air changes/hour

Difference between proposed and reference building

a. Solar Absorptance (α) of Walls = 0.6 & Roof = 0.7
b. Deemed to Satisfy (DTS) compliant insulation in all envelope elements (roof + ceiling, walls, floor)
c. DTS-compliant lighting and glazing (NCC glazing calculator 2014) to all orientations including roof lights

Table 3.   Energy consumed by building components for SRE construction in Sub-tropical climate.

A. Electricity breakdown

Room Electricity/ Computer + Equip (kWh/yr)	4750
Lighting/General Electricity (kWh/yr)	12666
Heating (Electricity) (kWh/yr)	250
Cooling (Electricity) (kWh/yr)	2201

B. Fabric and ventilation

Glazing (kWh/yr)	−1587
Walls (kWh/yr)	−1979
Ground Floors (kWh/yr)	−17524
Roofs (kWh/yr)	1204
External Infiltration (kWh/yr)	−2461
External Vent. (kWh/yr)	−795

C. Internal gains

Occupancy (kWh/yr)	2136
Solar Gains Exterior Windows (kWh/yr)	8074
Zone Sensible Heating (kWh/yr)	570
Zone Sensible Cooling (kWh/yr)	−4787
Total Cooling (kWh/yr)	−6823
Zone Heating (kWh/yr)	570

D. Energy (kWh/yr)

Annual Energy Consumption (AEC) kWh/yr	19867

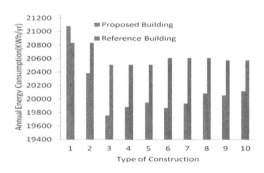

Figure 3.   AEC of the proposed office building for different wall constructions in sub-tropical climate.

Figure 3. Stabilised rammed earth (No 6) is the second best construction that can save AEC of the proposed building. However, the predicted Energy consumption of SRE with plasterboard lining (No 7) and Insulated Stabilised Rammed Earth, ISRE (No 8) constructions are slightly higher than heavy weight constructions (No 3, 4 and 5).

3.2 Tropical climate

In tropical climate of Australia, the scenario is similar to Sub-tropical climate, for 9 out of 10 constructions as have been found from the simulation results. These constructions demonstrated less energy consumptions than Reference Buildings' energy consumptions which satisfy the NCC Energy performance criteria. ISRE wall (No 8) construction showed lower energy consumption than any other constructions (Fig. 4). ISRE is a better energy performance indicator compared to heavy weight constructions (No 3, 4 and 5) and light weight constructions (No 1, 2 and 9). However, Brick veneer wall with added R1.5 insulation (No 10) showed almost same energy consumption compared to ISRE. The second best alternative construction after ISRE is to use SRE with plasterboard lining (No 7) which is a better performer than light weight walls (No 1 and 2) and heavy weight wall (No 5).

3.3 *Temperate climate*

In temperate climate of Australia, some of light weight (No 1 and 9) and heavy weight constructions (No 3, 4 and 5) are not complying with NCC 2014 for this proposed office building (Fig. 5). These constructions of the proposed building demonstrated higher energy consumption than Reference Buildings. The energy performance of ISRE (No 8) is better than any other constructions. However, SRE (No 6 and 7) are not complying with NCC 2014. To resolve this problem of rammed earth constructions, the change of other elements such as glazing can be a possible solution in this climate. Change of glazing Solar Heat Gain Coefficient (SHGC) such as low-e clear glass (U value 3.6, SHGC 0.68), instead of single clear glass with plasterboard lining to walls, can reduce the energy consumption to satisfy the criteria for section J Energy Efficiency (No 3–7: Solution for Proposed Building). Adding R1.5 insulation to walls (No 1 and 9: Solution for Proposed Building) can reduce the energy consumption of the proposed building. However, no insulation is required for SRE compared to Cavity panel (No 1) and Brick veneer wall (No 9) that satisfy the criteria of NCC 2014 (Fig. 5).

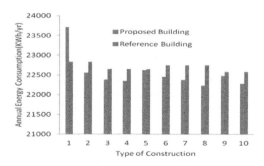

Figure 4. AEC of the proposed office building for different wall constructions in tropical climate.

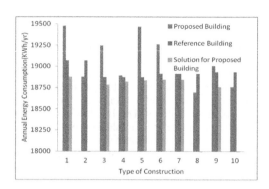

Figure 5. AEC of the proposed office building for different wall constructions in temperate climate.

4 CONCLUSION

Energy-efficient building designs and constructions are a mandatory requirement for building approval from local councils in Australia. To satisfy the energy efficiency requirement, for a specific construction, is a complicated process for building designers, consultants, contractors and researchers. Builders and contractors demand to reduce the construction cost and want to use minimum insulation in building fabrics with low cost glazing. These are the challenges to building designers and researchers. The results for the rammed earth constructions of single storey commercial office building from the above analysis in the Sub-tropical, tropical and temperate climate zones indicate that it can be an alternative option for low cost building construction that complies with energy efficiency requirement. Based on simulation results, it can be concluded that rammed earth walls performed better in Sub-tropical and tropical climate compared to temperate climate of Australia, as no changes to other elements are necessary; whereas, in temperate climate it may require some changes of other elements such as glazing to compensate the heating, cooling and overall annual energy consumption. The results for rammed earth constructions also predict the lower energy demand which may lead to less carbon emission for Australia. More research is required for structural stability, durability and life cycle analysis of the rammed earth wall construction for different types (e.g. Retail, School, etc.) of buildings in Australia. Australian government, building owners, operators, contractors, designers and researchers can work together to develop a sustainable technology plan for rammed earth constructions and encourage people to use this environment friendly construction.

REFERENCES

Allinson, D. & Hall, M. (2007). Investigating the optimisation of stabilised rammed earth materials for passive air conditioning in buildings, *International Symposium on Earthen Structures*, 109–112.

Building and Construction going green (2011) accessed from http://www.careerfaqs.com.au/news/news-and-views/building-and-construction-industry-going-green.

Ciancio, D. & Beckett, C. 2013. Rammed earth: an overview of a sustainable construction material. In Proceedings of *Third International Conference on Sustainable Constructions Materials and Technologies, 19-21 August,* 2013, Kyato, Japan.

CIE, Centre for International Economics (2007) retrieved from http://www.yourbuilding.org/library/carbonfootprint.pdf.

Dong, X., Soebarto, V. & Griffith, M. (2014) Strategies for reducing heating and cooling loads of un-insulated rammed earth wall houses, *Energy and Buildings*, 77; 323–331.

Energy Efficiency in Buildings (2009), World Business Council of Sustainable Development (WBCSD) accessed from http://www.epe-asso.org/even/91719_EEBReport_WEB.pdf.

Faure, X. & Le Roux, N. (2012). Time dependent flows in displacement ventilation considering the volume envelope heat transfers, *Building and Environment*, 50; 221–230.

Larsen, F.S., Filippín, C. & González, S. (2012). Study of the energy consumption of a massive free-running building in the Argentinean northwest through monitoring and thermal simulation, *Energy and Buildings*, 47; 341–352.

Orosa, J.A. & Oliveira, A.C. (2012). A field study on building inertia and its effects on indoor thermal environment, *Renewable Energy*, 37; 89–96.

Page, A., Moghtaderi, B., Alterman, D. & Hands, S. 2011. A study of the thermal performance of Australian housing, The Priority Research Centre for Energy, The University of Newcastle.

Section J, Energy Efficiency, *National Construction Code, Australia* (2014) accessed from http://bca.saiglobal.com.

Taylor, P., Fuller, R.J. & Luther, M.B. (2008) Energy use and thermal comfort in a rammed earth office building *Energy and Buildings*, 40 (5), 793–800.

Rammed Earth Construction – Ciancio & Beckett (Eds)
© 2015 Taylor & Francis Group, London, ISBN 978-1-138-02770-1

Earth Building—how does it rate?

P. Hickson
Director and Manager, Earth Building Solutions, Australia
President, Earth Building Association of Australia, Australia

ABSTRACT: This paper explores how Earth Building, including Rammed Earth, is placed to meet the challenge of the new millennium—reducing GHG emissions. The focus is on reducing GHG emissions in the built environment by the greatest amount and by the most effective and sustainable means.

The best approach to heating and cooling is about maximizing gains through harnessing natural conditions like solar gain, breezes and the cool of night in naturally conditioned ventilated buildings. The recommended design paradigm differs significantly from the sealed and insulated box model supported by the Energy Efficiency Provisions in the National Construction Code (NCC). That model concentrates on minimizing losses from conditioned spaces through the building envelope.

The paper uses a qualitative case study of an earth building to illustrate the difference between the two approaches. The predictive thermal performance, as measured by the Nationwide House Energy Rating Scheme (NatHERS), is compared with the actual as built performance assessed by the National Australian Built Environment Rating Scheme (NABERS) and a total Life Cycle Analysis assessed by eTool. The case study is compared with National averages and legislated standard benchmarks for new buildings in NSW and Australia. The paper reaffirms that climate responsive designed, mass-linked ventilated, earth buildings produce the best possible ecological, economic and social outcomes.

1 INTRODUCTION

Earth has been used as a building material for 11,000 years and still houses 1/3 to 1/2 of the world population. In wealthier countries in recent times it has been recognised the importance of earth as a building material, not in terms of providing shelter, but in terms of best sustainable practice. This trend exists across the world. Earth cab ne referred to as the ultimate green building material.

The purpose of this paper is to raise debate and awareness and find solutions around the impasse in gaining favourable assessments.

2 A SINGULAR FOCUS ON BUILDING ENVELOPE IS NOT THE ANSWER

The singular element that is receiving all of the attention, research and legislation in the area of building energy efficiency in Australia and elsewhere around the world is the external fabric of a building. Interest is focussed on the following questions: How well insulated, how well sealed against leaking air? What is the internal volume to external fabric surface area ratio? What is the heat flow coefficient and capacitance of materials used in floor, wall and roof elements? What are the size, orientation, specification, shading and value of fenestrations (glazed openings)?

The Building Code of Australia (BCA) J0.2 and BCA 3.12.0.1 state that, to reduce heating and cooling loads, a building must achieve an energy rating using house energy rating software. House energy rating software is defined in the BCA as follows: *"House energy rating software means software accredited under the Nationwide House Energy Rating Scheme and is limited to assessing the potential thermal efficiency of the dwelling envelope"*.

The great majority of post war housing in Australia uses lightweight construction and doesnt follow even basic climate design principles. Two-thirds of new homes are fitted with at least one refrigerated air-conditioner. Western Australia (WA) is the exception to this preference for lightweight building; the building stock is predominantly full brick masonry and rammed earth, though this is changing through the introduction of energy efficiency measures. So recently, lightweight construction is gaining a foothold in WA.

There exists a political imperative for energy efficiency improvements. Australia assesses the energy efficiency of proposed new buildings across the country using Australia's own House Energy Rating scheme (HER), NatHERS.

NatHERS is focussed on improving the performance of the external envelope of homes fitted with air conditioners, buildings mostly built without regard for simple climate responsive design principles and without "effective" mass. Why the focus on the envelope? Because poorly designed, effective lightweight project homes are the reality and the problem.

Once the proposed building plan is presented to the NatHERS assessment process, all that can be done to these buildings to make them more energy efficient is, seal them tightly, insulate them well and control ill placed windows with shading and/or double glazing. It is remedial action, not best practice. It is like spending the entire health budget on giving sick patients triple bypasses rather than spending something on preventative measures like promoting a healthy diet and exercise.

The NatHERS Software Accreditation Protocol states that in relation to building energy efficiency standards, the NatHERS accredited software must be used in Regulation Mode and operated in accordance with the 'Principles for Ratings in Regulation Mode' document.

According to Maria Kordjamshidi, HERs from around the world are unable to adequately model anything but sealed insulated conditioned buildings (Kordjamshidi 2011). As NatHERS doesn't even have basic ventilation logic, it is not capable of modelling anything but conditioned, well-sealed, well-insulated buildings. It doesn't model these buildings as they are operated in reality or as they need to be operated for the health and safety of occupants with minimum air changes. It doesnt allow for modelling using appropriate ventilation logic to maximise efficiency.

Unfortunately the problem is made worse because, due to its design limitations and Protocol, NatHERS is discouraging and disallowing naturally conditioned buildings with more effective energy efficiency outcomes. It is actively promoting poor building outcomes. This may be an unintentional but it is a real consequence.

3 A BETTER APPROACH IS NEEDED AND SHOULD BE ALLOWED AND SUPPORTED

It seems that major project builders constructing those ubiquitous lightweight buildings support NatHERS because they can use the same design in any climate on any site in any orientation and still easily comply with NatHERS.

Buildings proposed for tropical Darwin with a latitude just 9 degrees south of the equator should look totally different to buildings proposed for a much colder Hobart, 40 degrees south of the equator. Buildings proposed for a humid subtropical Sydney should be vastly different from those proposed for a Mediterranean climate of Adelaide or Perth even though they share similar latitudes. Many inland towns and cities like Alice Springs

have a desert climate with super low humidity and freezing cold to scorching hot conditions.

There are ancient solutions to be found in a rich variety in Vernacular Design (Steen and Komatsu 2003) and Australian Aboriginal vernacular buildings and settlements (Memmott 2007). What all of these approaches have in common is making the most of a particular climate and the principles have been trialled and tested over time until workable solutions were clearly evident and repeated and adopted as Vernacular Architecture. The varied designs were as remarkable as the variance in the climates. Many of these buildings especially those in warmer climates are naturally conditioned and often opened or permanently open to outside airflow. They are using natural systems like the heat of the day and solar access or shading, cool of the night, water and breezes. These natural conditions vary from place to place and within days, seasons and years.

The design needs to balance these natural conditions to provide the best outcome in terms of comfort to meet the expectation of occupants. All of them, even those in cold climates, were low carbon intensive. Most relied on carbon neutral biofuels like fire-wood for heating and didn't require cooling.

Designs varied dramatically with climate though earth was utilised in every climate. Often, this involved varying the density and thickness of earth walls. (Minke 2006) has produced some interesting data on the values and thermal characteristics of varying density. He has measured thermal values for earth walls ranging in densities ranging from 400 kg/m^3 to 2100 kg/m^3. Only values for mud brick and Rammed Earth are recognised with NatHERS. Buildings have taken advantage of the hygrothermal properties of earth walls. Hygrothermal properties been studied in laboratories by Minke (2006) and Allinson and Hall (2010) and have proven to be valuable in moderating humidity. One of the shortcomings of NatHERS is the fact that it cant model the hygrothermal behaviour of walls.

Mass is essential in moderating, balancing both temperature and humidity in naturally conditioned buildings. Mass is called **Fabric Energy Storage** (FES) and is being championed by an engineer Tom P. De Saulles in new energy efficient concrete buildings in the UK (de Saulles 2005). Professor Gary Baverstock of Ecotect Architects in Perth who specialises in Rammed Earth designs refers to the approach as **mass-linked ventilated buildings**. Engineers are now also using biomimicry in architecture to design better sustainable buildings. An example is the Eastgate shopping and office complex in Harare, Zimbabwe designed by Architect Mike Pearce in collaboration with Arup engineers (BG 2014). The building saves 90% of operational costs. Yemen

Skyscrapers provide evidence that these technologies and approaches are not new but have been used for hundreds if not thousands of years.

Every year the Earth's surface receives about 10 times as much energy from sunlight as is contained in all the known reserves of coal, oil, natural gas and uranium combined. This energy equals 15,000 times the world's annual consumption by humans. Yet our NCC Energy Efficiency Provisions do not require buildings to achieve a degree of "free running" which would encourage appropriate climate responsive design. To not collect this free energy and store it in thermal mass within the building during a winters day is plainly illogical (Birkeland 2008). To not collect and reuse the cool of the night in summer is equally nonsensical. To collect natural energy and allow it to flow through walls may be considered loss by some. But, if it is free energy or from a carbon neutral source, it is natural cycling of energy.

4 A CASE STUDY

So what tools are available to guide, promote and validate energy efficient design for building professionals and the client? There are many tools available to assess proposed buildings and existing buildings. Kordjamshidi (2011) explores all of the—HERS schemes from around the world and proposes a method that can be used to model free running climate responsive buildings. She concluded that—HERS cant model naturally conditioned buildings and the best approach is to promote these buildings and create a tool to assess it.

A 23-year Mud Brick (MB) house was subjected to a retrospective assessment using NatHERS to see how it performs against current legislation regarding energy efficiency. The assessor was Tony Isaacs, a well-known and respected energy assessor. Tony is a consultant to NatHERS and serves on the NatHERS Technical Advisory Committee.

AccuRate, FirstRate and Bers Pro are approved tools that can be used to conduct NatHERS assessments. AccuRate was used for this assesment (Isaacs 2014). The house was modelled in both regulation mode, used for official assessments, and in free running mode to describe the buildings behaviour if not heated or cooled and allowed to respond to the climatic conditions outside. For comparison, the same energy assessment was carried out for a modern Brick Veneer (BV) house nearby. Regulation Mode was used for the assessment accruing heating and cooling loads to maintain comfort with the building closed and simulation of annual external climatic conditions applied. The dwellings are both located in a cold climate area (Zone 6 as determined by NCC and Nowra, Climate Region 18 in NatHERS).

5 THE RESULTS

NatHERS stipulates that new buildings must achieve 6 Stars in energy efficiency. The result for this house was 3.2 Stars, as compared to 2.2 which is typical for houses of similar age (Isaacs 2014). The new typical BV project home rated 6 Stars though needed extensive remedial attention to improve energy efficiency due to poor design and orientation (Table 1). To achieve 6.0 Stars in this climate a total energy load of 81 MJ/m^2 is required.

5.1 Comments on results

The assessor remarked on the very low cooling load result for the MB house that is 3.33 times better than the modern BV house. It confirms the MB house can do without air conditioning though the BV house would be reliant or reasonably uncomfortable without it.

There is no doubt the old earth home specification could be improved if newly proposed and the design could be improved for solar gain. The ventilation logic of the MB house is not implemented in NatHERS. The 6 BV house might perform worse than assessed because ventilation will need to be used by the occupants to maintain healthy indoor air quality. Heat exchangers are rare in Australia because of own benign conditions. Achieving 1–2 air changes per hectare (ACH) in buildings, a recommended level to maintain high air quality (Jones 1999), would adversely impact energy efficiency outcomes in NatHERS by 40%. The energy use in the BV home may well be higher in reality than predicted because, according to Predicted Mean Vote (PMV) logic, people with air conditioners are more likely to set the temperature to provide conditioned PMV comfort settings rather than accept adaptive comfort. The adaptive comfort thermostat settings are only appropriate to naturally conditioned buildings according to the

Table 1. NatHERS results for MB and BV houses.

The MB house achieves a 3.2 Star rating

	MJ/m^2
Energy load	
Heating	177.2
Cooling	11.3
Total	183.5

The new BV project house achieves a 6.0 Star rating

	MJ/m^2
Energy load	
Midrule Heating	43.0
Cooling	37.6
Total	80.7

American Society for Heating, Refrigeration and Air-conditioning Engineers, ASHRAE. If the NatHERS protocol accepted heating thermostat settings were more appropriate for the MB house it wouldn't have needed anywhere as much energy because adaptive comfort is achieved at a much lower temperature. This would more accurately reflect our expectations of adaptive comfort.

AccuRate tool as used within NatHERS in its current form offers an understanding of the predicted thermal performance of an MB house as if it were a conditioned, sealed and insulated box. A 3.2 star rating proves it is not a particularly good example. NatHERS could be adapted to model naturally conditioned buildings in free running mode using varied ventilation logic. A Star Rating could be based upon percentage of time in the year that an adaptive level of comfort was achieved without energy and then considering the energy required and carbon intensity of energy to supplement natural conditioning. If 10 Star buildings were autonomous in heating and cooling then a building that achieves adaptive comfort for 75% of a year in free running mode that is without heating or cooling it would be assessed 7.5 Stars.

6 CONCLUSIONS

We shouldn't be focusing all of our attention on what is simply remedial action to a flawed model. Unless—HERS tools are enabled to model and properly assess naturally conditioned buildings they will not be optimizing design but instead risk encouraging poor design. And they will continue having a negative impact on the evolution towards buildings not merely energy efficient but more autonomous in space heating and cooling.

Achieving small incremental improvements in energy efficiency by simply sealing and insulating building seeks to minimize losses in poor buildings and doesn't get us anywhere towards our international obligations, sustainability benchmarks or goals. We need to shift the paradigm to positive development where buildings are not a constant drain on resources and energy but add to both ecological and social capital. We need to start designing naturally conditioned buildings (call them adaptive designs, climate responsive designs, vernacular design, bioclimatic designs etc.), aiming at maximizing gains not minimizing losses as a first principle. These buildings need to be low embodied energy, energy efficient, durable, low maintenance, safe, desirable, affordable, comfortable and offer healthy indoor air quality. The best way to achieving this is to combine the lessons of vernacular buildings essentially the crystallization of 10,000 years of trial and error, happy accidents and experimentation then utilise the powerful tools and computer technology of the present to optimize design. We will then begin to design and construct the sustainable buildings of the future from the lessons of the past. There is no doubt Earth—the ultimate green building material—has a part to play and this.

REFERENCES

Allinson, D. & M. Hall (2010). Hygrothermal analysis of a stabilised rammed earth test building in the uk. *Energy and Buildings 42*, 845–852.

BG (2014). Architecture. [accessed: 30/10/2014].

Birkeland, J. (2008). *Positive development: From vicious circles to virtuous cycles through built environment design*. Routledge.

de Saulles, T. (2005). Thermal mass—A concrete solution for the changing climate. Technical report, The Concrete Centre, Surrey (UK).

Isaacs, T. (2014). Rating report and analysis for Lot 6 Island Point Road, St Georges Basin. Technical report, NatHERS.

Jones, A. (1999). Indoor air quality and health. *Atmospheric Environment 33*(28), 4535–4564.

Kordjamshidi, M. (2011). *House rating schemes: From energy to comfort base*. Springer.

Memmott, P. (2007). *Gunyah, Goondie and Wurley: The Aboriginal architecture of Australia*. University of Queensland Press.

Minke, G. (2006). *Building with earth—Design and technology of a sustainable architecture*. Birkh auser, Basel.

Steen, B. & E. Komatsu (2003). *Built by hand—Vernacular buildings around the world*. Gibbs Smith.

Rammed Earth Construction – Ciancio & Beckett (Eds)

Rammed Earth in a concrete world

M. Krayenhoff
Director, Tech Energy Systems, SIREWALL Inc., Salt Spring Island, BC, Canada

ABSTRACT: Now that SIREWALL (Structural Insulated Rammed Earth) Rammed Earth (RE) is typically stronger than generic concrete and that strength is achieved while using less than 10% cement, more commercial RE building opportunities are available. Having RE in commercial buildings creates credibility for everyone in the RE field. The public learns that this product technology is accepted by architects, clients, and general contractors. What the public doesn't see is the difficulty that RE has fitting into the commercial construction paradigm. Two of the biggest sources of that difficulty are the expectation that RE is just like concrete, only it looks different, or sometimes it is viewed as a decorative element, like wallpaper. Neither of these widely held perceptions is accurate or useful.

1 INTRODUCTION

There are some strong headwinds that will be faced by any Rammed Earth Subcontractor (RES) when embarking on commercial work. General Contractors (GC) are risk averse and committed to their system of getting the job done. Introducing a new building technology like rammed earth introduces perceived risk and upsets their system. The rammed earth industry needs to develop a way to ease into the established commercial construction paradigm. All emerging technologies face integration issues and only by exposing the difficulties and discussing them, can we begin to change the dynamics. This paper begins that conversation.

2 INTEGRATION ISSUES

Building systems for commercial buildings involve a limited palette of structural materials (primarily concrete and steel). Once the structure is in place there is a much broader range of materials that are used as veneers to pretty the building up. That broad list would include such things as claddings, paint, floor coverings, drop ceilings, etc. The end result is that very seldom does one experience the structure, other than exposed concrete columns. Wall assemblies remain a structural mystery to the viewer. With SIREWALL, 18–24" thick RE walls with a 4" layer of insulation hidden within, the opposite is true (Figure 1). The structure is fully exposed on the inside and outside. The viewer knows immediately what it is made of. There are strong emotional and environmental benefits to that nakedness.

Achieving the naked wall requires more protection through the construction process than a

Figure 1. Loadbearing 51' tall SIREWALL.

concrete wall that will be clad. The naked wall takes longer to build, but when done it requires no cladding. The sequencing of construction is altered. It takes longer to get the roof on, but the finishing takes less time.

In northern climates, the fair weather window is ≈7 months. Working through the winter will add to the cost significantly. A Midwestern construction adage is that an hour's work in the summer takes two in the winter. On top of that is the cost of keeping soil, mixing machine, formwork, and curing walls warm. Winter work is doable, often unavoidable, and a fact of life in northern climates.

Inside the pressures of commercial construction, the introduction of an unknown variable like rammed earth may cause the GC to act less than graciously. He typically knows almost nothing about rammed earth, and is charged with making sure it all goes well such that the client and architect sign off. It is an uncomfortable position to be in as he is charged with directing someone who knows so much more than him. Most often the GC does not take advantage of the RES's knowledge and experience, and instead feels threatened by it.

When the GC hears that SIREWALL has similar strengths to concrete (although with less than 10% cement, for example the wall shown in Figure 2 with a compressive strength of 6670 psi (46 MPa)) his first instinct is to classify SIREWALL as coloured concrete. This gives him comfort as he thinks he knows what to do, having handled many concrete subtrades. That is the root of the problem, and from that moment forward RES will have to deal with the fallout of that erroneous expectation.

Typically there is a honeymoon period in the project where the GC and architect are enamored with the RES's capabilities and knowledge. Occasionally the good vibes last till the project is complete. However it is more typical and unfortunate that the honeymoon ends before the project does, due to financial/scheduling pressures and erroneous expectations.

The commercial construction paradigm is conflicting in nature. The GC wants as few change orders as possible and wants to pay as little on each one as possible. Subtrades want the opposite. That friction is normally tempered by the "future shadow". Both parties may need each other in the future and don't want their future prospects diminished by full out conflict, or their local reputation tarnished. For the RES working away from home, there is no future shadow. Extreme grinding and unfair practices do not hurt the GC and can benefit their bottom line. This is a critical distinction for the RES to recognize. Working at a distance from your home community, you lose the home field advantage that ensures a measure of fair play.

Most projects have many points of friction due to erroneous expectations, and on some only one or two have surfaced. The point is that these issues have costs attached that have to be priced into the initial bid. At the time of bidding it is impossible

Figure 2. RE wall with 6670 psi (46 MPa) compressive strength.

to determine the number and cost of issues. Failing to price in these issues, can cause financial stress or bankruptcy (not at all uncommon for RE subtrades moving into commercial work).

Pricing in these issues makes the RE product not as competitive as it might be. The Client is paying for the cost of unnecessary friction due to lack of future shadow and the expectation of full out grinding. The Client is also paying the GC an inflated amount for the GC to handle the anticipated difficulties of the RE subtrade. In total the Client is paying 50% to 100% extra due to concerns that the RES and GC have about working together.

3 INTEGRATION ISSUE SOLUTIONS

So what could RES do at the outset to avoid or minimize these issues?

SIREWALL support: Ideally, the architect and engineer support should take place prior to the GC being involved. We have seen this work well and recommend it highly. Unfortunately, many projects switch to SIREWALL late in the design process, just before tender. In such a circumstance, the Architect needs to step in to make it clear to the GC that he/she will need direct communication with SIREWALL/RES prior to the build and throughout the build. The best option is round table collaboration between Architect, GC, and SIREWALL/RES. **Feasibility:** Prospecting, mix design, strength and colour: Press to launch this work earlier than the minimum one month out, as delays here can cause cascading scheduling issues.

SIREWALL expectation management: A manual for the GC with what to expect in the SIREWALL building process. Managing expectations at the outset includes the recognition that local soils are unique and are reflected in the finished product, and that SIRE WALL is an artisanal product.

Payment structuring: Given that there is an inherent conflict between the GC and RES interests, the best option is to have the Client pay RES directly, as they would any other artist. This avoids the future shadow issue and ensures the Client is getting the best value for dollar spent on RE.

4 CONCRETE COMPRESSION TEST RESULT VARIABILITY

Because concrete is so ubiquitous and well established in our building culture, it is inevitable that RE will be compared, at every turn, with concrete.

Engineers ask for compressive testing of concrete and RE in order to gain comfort that what is being produced will meet the structural requirements that they have calculated.

It is important to distinguish between the strength of the samples and the strength of the wall. They will not be the same, as they are created in different conditions. For concrete some differences are:

Consolidation uniformity—ensuring there are no voids in a concrete sample is a relatively simple matter. Vibrating concrete in the wall with rebar produces inconsistent results (Neville 2011). *The sample will be better than the wall.*

Particle size distribution—there is no drop height when making cylinders, whereas in concrete a drop height of 15'–20' (4.6 m–6.1 m) or more is common. Dropping from height is known to create uneven aggregate distribution in the finished wall (Roussel 2011). *The sample will be better than the wall.*

Water/cement ratio—although it is frowned upon, it is fairly common for water to be added to the mix if it is getting too stiff for the pump truck. Samples taken early in the pour do not represent the inflated water/cement ratio that may occur later in the pour (Neville 2011). *The sample will be better than the wall.*

Quality control—on a 15'–20' tall concrete pour, there is almost no visual quality control possible at the bottom of the forms. The concrete gets poured in and vibrated with fingers crossed. Yes there are protocols regarding vibration, but if something is not going well there is no way to know at the time of the pour. *The sample will be better than the wall.*

5 RAMMED EARTH COMPRESSIVE STRENGTH VARIABILITY

For RE, the mix design, the process of consolidation and the delivery is far different than with concrete. Commercial RE could be mixed in a custom volumetric mixer, delivered with a crane/hopper / elephant's trunk, and consolidated with pneumatic rammers.

In mix design using larger stones adds strength, as long as the stone size does not exceed 1/8 of the wall thickness. For a 24" (600 mm) solid SIREWALL our mix on a recent project had 2.5" (64 mm) stones in it. Consolidation is achieved by rammers that apply 1,000 blows per minute to the damp soil. The person ramming is in direct visual and tactile contact with the damp soil that he is compacting into stone. Delivery through the elephants trunk allows the soil to be placed where it should be with visual confirmation for quality control. Engineered sandstone samples are rammed in 6" (150 mm) diameter cylinders. The process is difficult for the most experienced rammer, especially near the top of the cylinder.

Consolidation uniformity—rammers go up and down 1,000 times per minute and have a 4"

(100 mm) stroke. It is very difficult to get the top of the cylinder rammed to the same degree as the middle and bottom. Ramming inside a 24" (600 mm) wall is so much easier and thorough than doing test cylinders. *The wall will be better than the sample.*

Particle size distribution—the delivery of the material is so much more controlled than with concrete. There may be a 2.5" (64 mm) stone in one of the cylinders, setting up a shear plane failure in the testing (Bryan 1988). In the wall there is no downside to the occasional large stone, just upside. *The wall will be better than the sample.*

Water/cement ratio—this is tightly controlled by the volumetric mixer and it is impossible (and unnecessary) to add water at the wall. This will be the same for the samples as for the wall. *The wall will be equal to the sample.*

Quality Control—in the wall the person ramming is delivering the material and has good visual, tactile, and auditory appraisal as to the quality of the consolidation. In the cylinders there is not sufficient sample size to gain benefit from those signals. *The wall will be better than the sample.*

In all cases the concrete wall is not as good as the sample. With RE, in all cases the RE wall is equal to or better than the sample!

In many countries the concrete in the wall is generally expected to reach 75–80% of the sample strength. With RE, the multiplier has as yet to be determined but will almost certainly be over 100%. We do know that the scale of the sample is not large enough to be representative of the RE wall. We do know that the concrete samples will break in a consistent manner, whereas the RE samples break in a variety of ways. We attribute that to shear planes in the sample. We do know from a recent project that 14 samples crushed at 3 days yielded strengths between 1070 psi (7.4 MPa) and 1910 psi (13.2 MPa) and the average was 1540 psi (10.6 MPa). RE compressive strengths are plus or minus 25% from the average. We don't know what the multiplier for actual wall strength is. It seems reasonable to propose that, based on the above variables, that the wall strength will be higher than the average. If the argument is valid that the engineered sandstone sample is the worst case, then we might take the highest compressive strength (1910 psi or 13.2 MPa) as representing the wall strength. It seems reasonable to go forward with using the average (1540 psi or 10.6 MPa), the highest (1910 psi or 13.2 MPa), or midway (1725 psi or 11.9 MPa) as representing the wall strength at 3 days.

It is appreciated that there is a desire for consistency in cylinder outcomes. To get better consistency it is possible to screen out all the big stones to prevent shear planes in the sample. This would however not accurately represent the strength that is gained from having larger fractions in the mix.

The cylinders would then reflect even less strength than the wall. Due to the nature of how the engineered sandstone is put together, there will always be more variability than concrete.

6 RILLING AND BONINESS

Another significant and related difference between engineered sandstone and concrete is "rilling". Rilling is when the larger particles run down the angle of repose to gather at the bottom of a pile of soil. That pile can be soil in the gravel pit, a soil storage pile at the site prior to mixing, a soil pile that has just been mixed and delivered to the wall, or a soil pile about to be rammed as a sample. In each of those four circumstances a degree of inconsistency is introduced.

At the pit, a lack of awareness regarding rilling can result in misleading particle size distribution outcomes. Rilling in on-site soil piles can result in uneven mixes, although the mixing process typically handles that variable. Rilling in delivery can result in visually boney areas that may not please the eye (but have little impact on strength). Rilling in sample preparation (larger particles roll to perimeter of cylinder when soil is placed in cylinder) is very difficult to avoid, and it has very large impacts on compressive test results.

Boniness is the surface result when rilling has taken place. Typically it runs horizontally and is less than one inch in depth. The GC will see it as identical to "honeycomb" in concrete, where the uneven consolidation runs the full thickness of the wall and is a serious problem. Boniness may look like honeycomb, but is not a structural concern or failing. It is simply a surface visual condition. All RE walls will have some boniness. How much depends on the soil mix design which is dependent on the strength requirements and the soils available in the local area. Boniness is also a result of the delivery technique.

7 QUALITY CONTROL

Strength: Contractors should ram a cylinder per day and the engineer can crush as many as he likes to prove that the compressive strength is sufficient. **Art:** every wall is different as we use local soils that have unique qualities. It is impossible to create one wall identical to another wall. It is even more difficult to create a wall that will consistently match a site built sample wall. Inside of the rammed earth reality that all rammed earth walls are inconsistent in colour and texture, and may exhibit some non-structural cracking or efflorescence, the best approach is to select an existing SIREWALL building as a reference standard. That reference standard will have a certain quality of forming, of form lines, of cold joints, of reflectivity consistency, of lift height consistency, of panel size, of patching, and of top of wall finish quality. The new wall will most likely be better than the reference wall in some areas and not so good in others. The overall quality should be equal within reason.

8 CONCLUSIONS

In conclusion, rammed earth is beginning to find its place in the commercial world but the RES still needs to travel to find sufficient work. In "working away" the RES will need to establish relationships and manage expectations of GCs who have never before worked with RE. While compressive strength is the key to broader adoption of rammed earth, the common testing techniques produce variable results and the interpretation of those results needs more research. Rilling is not found in concrete but definitely impacts the entire production chain for Rammed Earth. Boniness is a result of rilling and is not a structural issue. Wall quality needs to be determined relative to the rammed earth industry, not the concrete industry. It is important that we begin the conversation about integrating the emerging rammed earth industry into the existing commercial paradigm.

REFERENCES

Bryan, A.J. (1988). Criteria for the suitability of soil for cement stabilization. *Building and Environment 23*(4), 309–319.

Neville, A.M. (2011). *Properties of concrete* (5th Edition ed.). Pearson Education Ltd, Essex (UK).

Roussel, N. (Ed.) (2011). *Understanding the rheology of concrete*. Elsevier.

Rammed Earth Construction – Ciancio & Beckett (Eds)
© 2015 Sirewall, Salt Spring Island, BC, Canada, ISBN 978-1-138-02770-1

Rammed earth thermodynamics

M. Krayenhoff
Director, Tech Energy Systems, SIREWALL Inc., Salt Spring Island, BC, Canada

ABSTRACT: Building standards regarding energy efficiency continue to rise around the world in response to increasingly expensive energy and climate change. Unless energy becomes cheaper and climate change stops, the energy efficient building trajectory will continue to have a big impact on design. Energy efficient detailing typically drops between the architect and the engineer. Both make an effort but neither is well trained in evaluating and designing for a new wall assembly such as insulated (or uninsulated) rammed earth. Shifting the emphasis from supply side technologies to reducing the demand for energy is a vital and necessary response to climate change. Insulated rammed earth excels in reducing energy demands. However that excellence is only possible when 8 common thermodynamic flaws are understood and avoided. The growth and reputation of the rammed earth industry and particularly the insulated rammed earth industry will be largely determined by how well thermal envelopes are designed and built.

1 INTRODUCTION

The US Green Building Council states that 41% of the total energy consumed in the US is by buildings. Of that 41%, half (20%) is used for heating and cooling of buildings. That 20% of total energy demand is a direct result of the quality (or lack of quality) of the thermal envelope (USGBC 2014).

Addressing energy demand from buildings is either done on the supply side (coal, gas, wind, hydro, nuclear) or the demand side (targeting net zero buildings). Increasing the supply of energy is good for a countrys GDP and a handful of companies, but not much else. Focussing instead on reducing demand is good for the environment and our long term economic wellbeing. Changing suppliers or adding capacity to solve the energy crisis is the same thinking that got us into it. Climate change will bring forth increasingly severe challenges to the planets housing stock. Extreme temperatures, wind, flooding, wildfires, and insects, will require buildings to be more durable and energy efficient.

It seems inevitable that someday soon all new buildings will be required to be net zero and have durability measured in centuries. As well, local economies are moving toward requiring local buildings are made with local materials and local labour. The emerging rammed earth industry, and particularly the insulated rammed earth industry is well positioned to fill that seemingly inevitable future demand.

2 TO INSULATE OR NOT

Buildings built today will be facing a future with an increased range of temperatures. Todays standard deviations in temperature will become a fond memory as polar vortexes and heat waves become normal. Design requirements have traditionally looked at past weather patterns and ensured that comfort could be maintained inside historic norms. That approach already no longer works.

The thermal flywheel effect of abundant uninsulated mass is ideally suited to address short term temperature swings. But what if there is a month long polar vortex or heat wave? In that circumstance having only mass will ensure the building is too cold or too hot. Only in locations where uncomfortable temperatures are projected to never be lengthy, does it make sense to not insulate. Those locations are few. For most of the planet it makes sense to use insulated rammed earth (see Figure 1).

There are many ways to do this. We tried three techniques before settling on the hidden plane of rigid board insulation in the middle of the wall. First we tried mixing zonolite into a wet rammed earth mix. The R value and strength were acceptable but not stellar. Then we tried insulation on the outside of the wall with stucco to protect it. That worked well thermally and structurally but we lost the visual appeal of the rammed earth on the exterior. Third, we tried forming a void in the middle of the wall. When ramming was done, the outside forms were removed and then the cavity forms were removed from the inside. That void in the center of the wall allowed easy access for the electrical and plumbing to take place and when that work was complete, insulation was poured in to provide the R value. What we settled on as the best was rigid board insulation hidden in the middle of the wall. That approach has been replicated by many around the world. It is now seen as the "obvious" way to insulate a rammed earth wall.

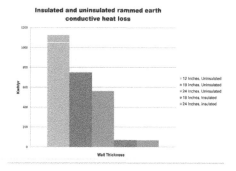

Figure 1. Changes in conductive heat loss with changing thermal mass and insulation.

3 INSULATED RAMMED EARTH FLAWS

Inside that approach there are eight detail flaws that can reduce the effectiveness and integrity of the thermal envelope. Flaws that undermine the effectiveness of insulated rammed earth are listed from maximum impact to least impact:

a) **The solid full width concrete bond beam—** that goes on top of the wall all the way around the building (Figure 2, left). Engineers like this as it works like a double top plate in framing. The rammed earth is sandwiched between two horizontal concrete elements. Concrete at 0.1 per inch over a 24" wall will have only R2.4 over its entire surface area. This detail will ensure more heat loss through the top of the wall than through all the windows and doors put together. Having a solid concrete bond beam largely negates the usefulness of insulating the wall in the first place.

b) **Windows and doors not placed in plane with insulation** in the extreme condition the windows and doors are placed as far as possible to the exterior of the wall, leaving at least 6" of the exterior wythe exposed to the interior environment (Fig. 3, upper). In addition to much heat loss will be condensation on the interior of this exterior wythe.

c) **Solid 3" RE around window and door openings** by not bringing the foam to the wall openings, a thermal bridge is created around each opening (Figure 3, lower). The impact of that seemingly in-significant detail increases the heat loss of the insulated rammed earth by 462% (Hall et al. 2012)! Condensation will occur in this condition as well.

d) **Suspended concrete slabs that extend to the outer wythe** second floor and roof slabs can be supported entirely on the inner wythe such that the thermal envelope is uninterrupted (Figure 2, right). However, running the concrete slab through to the outside wythe creates a thermal bridge similar to the solid concrete bond beam. There is a possibility of condensation with this condition.

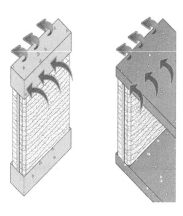

Figure 2. Concrete bond beam (left) and suspended concrete slabs (right).

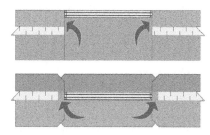

Figure 3. Out-of-line (upper) and in-line (lower) windows.

e) **Thermal envelope from top of SIREWALL is discontinuous** wood framing is much thinner in section and is often placed on the outer wythe such that the exterior of the woodframe finish is in plane with the exterior wythe of the SIREWALL (Figure 4). The thermal envelope of the woodframe is not directly on top of the thermal envelope of the SIREWALL, creating a discontinuity or thermal bridge. Also, roof framing on top of the SIREWALL needs to ensure thermal envelope continuity. If poorly executed, there may be condensation issues.

f) **Heat loss under the inner wythe** this can be significant or not depending on the distance that the heat needs to travel to pass from the indoor temperature to the outdoor temperature. There are two worst case scenarios;

i) the insulated rammed earth wall sits on a concrete slab on grade and the wall is backfilled only a couple of inches (Figure 5, left). The heat from the indoors travels from the bottom of the inside wythe under the insulation, through the footing, and up the outside wythe to the outside environment. The distance could be as little as 15" of R0.2/inch for a thermal bridge around the perimeter of the building of R3. This is not quite as bad as the bond beam but quite significant.

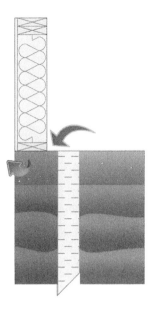

Figure 4. Discontinuous thermal envelope.

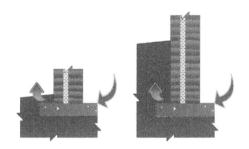

Figure 5. Shallow (left) and deep (right) backfill.

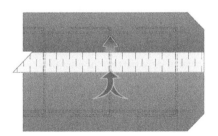

Figure 6. Heat loss across interwythe connectors.

ii) in circumstances where the soil is dry and the insulated wall is sitting on a footing that is well below grade (Figure 5, right), then the heat loss calculation is a function of the temperature of the soil at depth, not the temperature of the air. In cold climates, the temperature of the soil is normally far less extreme than the temperature of the air in winter so the

heat loss is far less. (In hot climates that connection to the soil through the inner wythe is a valuable cooling resource.)

g) **Poor perimeter drainage** moist backfill and footings will wick the heat away more quickly.

h) **Heat loss across the interwythe connectors (IWCs)** the 10M steel rebar that is typically used for IWCs is a good conductor of heat, but there is very little cross sectional area (Figure 6). A typical 24″ × 24″ grid of IWCs will reduce the overall R value of the insulated rammed earth wall by R1. This small thermal bridge can be virtually eliminated by using fibreglass or basalt rebar. No condensation issues.

The thermal envelope containing abundant thermal mass is a very low tech way to store heat and normalize indoor temperatures. There are no moving parts and it will work on Day 10,000 the way it worked on Day 1 with no maintenance in between. In a durable SIREWALL building its important to look at which flaws will be tolerated.

4 THE RADIANT ENVIRONMENT

In 2001, we used a remote thermometer in the 25' tall room shown in Figure 7. To our surprise the surface temperature of the walls, ceiling, and floor were all within 1 degC. We expected stratification. After testing other SIREWALL buildings, we now better understand the radiant environment created by high mass contained within high insulation that is punctured only with low e glazing. How it works is that the infrared energy, which is constantly bouncing around the space trying to equalize surface temperatures, now has no escape through windows (due to low e) and is stored inside significant insulated

Figure 7. Instrumented room.

thermal mass. The benefit is that stratification is eliminated and the human body is heated by the surfaces around it, not by the temperature of the air. Comfort is achieved at a much lower air temperature (eg., 16 degC feels like 20 degC). Its like being in a low temperature oven, which can be set to whatever temperature is most comfortable.

5 DATA LOGGED PERFORMANCE

A 2007 BCIT study looked at the performance of an unheated and unoccupied SIREWALL home over a one month period. Figure 8 looks at thermal performance over a month in Spring.

This house was built in 2002 and had Flaws (d), (e), and (f). Despite that, the performance shows an average outdoor temperature of 7°C (red line) and an average indoor temperature of 16°C (feels like 20°C). Thats 9°C of free heat (feels like 13°C) and temperature stability. Without the flaws the result would be even better.

The same BCIT data logger tracked the humidity over the month (Fig. 9). The outside variations go up and down daily (red line). Inside, the humidity is very stable, right in the middle of the 40%–65% human comfort zone. Possibly more important is that no mold can grow at less than 65%, and the hygrothermic capability of the SIREWALL to prohibit high humidity is clearly shown by the tiny variations in the graph (blue line). With enough SIREWALL surface area, it becomes nearly impossible for mold to grow in the building.

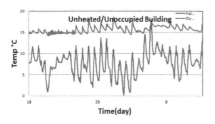

Figure 8. Internal and external temperature changes over monitored period.

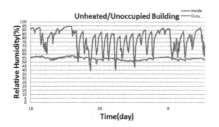

Figure 9. Internal and external humidity changes over monitored period.

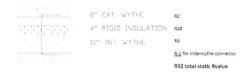

SIREWALL DETAIL

Figure 10. SIREWALL detail.

6 STATIC AND DYNAMIC R VALUES

Historically, building walls have been either mass with little insulation (eg., stone, concrete, and brick) or insulation with little mass (eg., straw bale or wood frame with insulation). Only recently are there walls with high interior mass, inside the thermal envelope, that are exposed to passive solar energy. The energy efficiency benefit of exposed mass contained within high insulation is expressed as the Dynamic R-value.

Jan Kosny's work on Dynamic R-values at Oakridge National Laboratory shows that the benefit is site dependent. Typical dynamic R-values are 1.5 to 2.25 times the Static R-value. Exposed interior mass, that is insulated, makes your wall at least 50% more energy efficient. Sites with more solar gain will be at the top end of that multiplier (225% more efficient). There is more detail on Dynamic R-values on page 567–570 in Modern Earth Buildings (Hall et al. 2012). Based on this research, SIREWALLs dynamic Rvalues are R48 to R72.

7 CONCLUSION

In summary, as we move forward into a climate changed environment, our buildings will need to be more energy efficient and durable than ever before. Insulated rammed earth buildings have much to offer this future. The reputation that they develop will depend in large part on the attention to detail in design and construction of the thermal envelope.

REFERENCES

Hall, M.R., R. Lindsay, & M. Krayenhoff (Eds.) (2012). *Modern earth buildings* (First Edition ed.). Woodhead Publishing, Cambridge, UK.
USGBC (2014). Green building facts. [accessed: 14/10/2014].

Rammed Earth Construction – Ciancio & Beckett (Eds)
© *2015 Taylor & Francis Group, London, ISBN 978-1-138-02770-1*

Thermal performance summary of four rammed earth walls in Canadian climates

C. MacDougall
Queen's University, Kingston, Ontario, Canada

K.J. Dick
University of Manitoba, Winnipeg, Manitoba, Canada

T.J. Krahn
Building Alternatives Inc., Codrington, Ontario, Canada

T. Wong
Stone's Throw Design, Toronto, Ontario, Canada

S. Cook
Aerecura Rammed Earth Builders, Castleton, Ontario, Canada

M. Allen
Muskoka Sustainable Builders, Huntsville, Ontario, Canada

G. Leskien
Zon Engineering, Kitchener, Ontario, Canada

ABSTRACT: Rammed earth construction is experiencing a renaissance in Canada, as in many parts of the world. The climate in parts of Canada is considerably colder than other countries and regions where rammed earth has been used more extensively throughout history. A straw bale insulated rammed earth greenhouse in Manitoba, two single family stabilized insulated rammed earth dwellings in Ontario, and a straw bale insulated compressed earth block building in Ontario are described and their energy performance is quantified against recorded climatic conditions. Pre-construction energy modeling is compared with actual performance, and the effect of thermal mass is discussed. Energy modeling and ongoing calibration of energy models through the refinement of thermal and hygrothermic parameters are briefly discussed.

1 INTRODUCTION

Globally, there has been renewed interest in various "natural" building materials and techniques. Some of these are very ancient, such as adobe, and others more recent, such as straw bale construction. In Canada, the use of earth-based construction has tended to be avoided, as it is seen as suitable for hot, arid climates. However, in recent years, there have been several buildings constructed in Canada that incorporate either rammed earth or compressed earth blocks into their structure. The first permitted, multi-wythe insulated rammed earth house in Ontario was constructed in 2012 by Aerecura Rammed Earth Builders. The insulated walls consist of inner and outer wythes of 150 mm rammed earth sandwiching 150 mm of polyisocyanurate insulation (Figs. 1(a) & (b)).

Three further examples are described in this paper. The first is in located in Winnipeg, Manitoba, at the eastern edge of the Canadian Prairies. The central Manitoba climate has average winter lows of minus 20 °C and average summer highs of plus 25 °C. The other three are located in southeastern Ontario, which has typical average winter lows of minus 10 °C and average summer highs of plus 27 °C. Selected monitoring data from these structures is discussed. This is, to the authors' knowledge, the first quantified evidence of the thermal performance of earth-based construction in Canadian climates.

Figure 1(a). Castleton, Ontario, rammed earth residence.

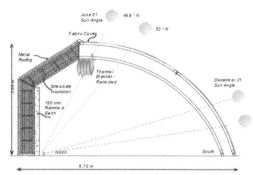

Figure 2. Cross section of greenhouse.

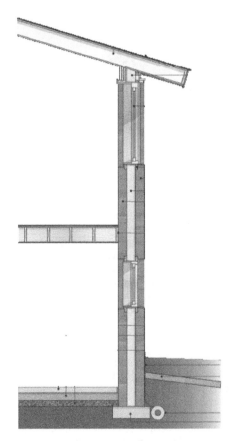

Figure 1(b). Typical wall cross-section, Castleton house.

2 PROJECT DETAILS AND MONITORING

2.1 *Alternative village greenhouse*

The Alternative Village is a research facility located on the campus of the University of Manitoba, Winnipeg, Manitoba. In 2011, a greenhouse (Fig. 2) was constructed at the Alternative Village as part of a research project investigating alternative food production techniques for northern communities. Wheat straw bales (460 mm deep, 400 mm high, 1066 mm long) were used for insulation behind a 150 mm thick rammed earth thermal storage wall. The north side was sheathed with galvanized metal panels fastened to 38×89 mm dimensional lumber girts and purlins. Fig. 3 shows the non-woven geo-textile used for the rammed earth. A well-graded granular material with a maximum grain size of 16 mm was dry mixed with 4% Portland cement by weight. The stabilized soil mix was placed in lifts not exceeding 150 mm in depth and then compacted between the straw insulation and fabric. Thermo-couples were installed at the fabric face, midway through the layer and at the interface between the straw and rammed earth.

Temperature (through 54 thermocouples) and relative humidity (through 6 Honeywell HIH-4000 relative humidity sensors) were monitored through the back wall assembly in addition to ambient indoor and outdoor conditions. Temperature was monitored through the rammed earth in three locations: the interface between the geotextile and soil, centre of wall and inter-face between straw and soil.

Fig. 4 is the temperature profile at the mid-height and mid-length location of the rammed earth wall. Data was collected from January through May of 2012, although only February to March is shown. The data illustrates the storage capacity of the rammed earth mass. These data were recorded when the thermal blanket was not in use, thus the only resistance to heat flow during non-sun hours was the greenhouse cover. The rammed earth at the strawbale inter-face was able to maintain tem-peratures between $-2°C$ to $+10°C$. On the coldest days during this time period the temperature at the inside surface of the wall was approximately 16°C warmer during nighttime hours than the ambient outdoor. The effect of the black, non-woven geo-textile is evidenced in the data by the substantial

Figure 3. Rammed earth back wall.

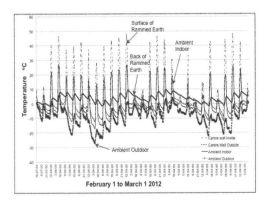

Figure 4. Temperature profile through rammed earth at mid-length, mid-height.

temperature difference. Over this period there were days when the surface temperature was approximately 48°C warmer than the outdoor ambient conditions.

2.2 Compressed earth block residence

The second building is two storey structure constructed north of Cobourg, Ontario (Fig. 5). The building has a usable floor area (measured to the inside of the walls) of roughly 84.5 m² (910 ft²) and an interior volume of approximately 217 m³ (7,660 ft³). The above-grade wall assemblies are a hybrid construction of Compressed Earth Blocks (CEBs) and straw-bales.

Heating was provided during the study period by radiant floor heating (ground floor only). Temperature and relative humidity readings were collected at four locations around the building. A multi-day analysis was carried out and involved "co-heating" and "cool-down" periods (Fig. 6a). During the two-day "co-heating" period, the building was maintained at a constant interior temperature

Figure 5(a). Southwest elevation of demonstration building.

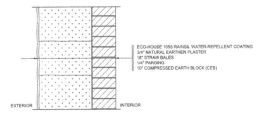

Figure 5(b). Typical wall section.

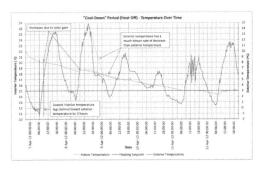

Figure 6(a). Heat loss/gain characteristics.

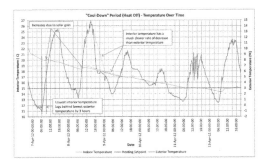

Figure 6(b). Cool-down period.

of 22°C. Immediately following the "co-heating" period, the set-point of the temperature controller reduced to 15°C, allowing the building to drift with the exterior temperature and effects of solar insolation. The graph of the "cool-down" period (Fig. 6b) indicates that on April 7 and April 8 a small increase in interior temperature (approximately 0.5°C) were recorded just before noon. From April 9–11 no such increase was observed. Weather station data for these days indicated that April 7 and 8 were clear with high solar insolation, while April 9–11 were overcast. These observations suggest that the demonstration home has a small response to solar heat. Also evident from the "cool-down" period is the difference in the rate of temperature decrease inside the building when compared to that of exterior temperature as it tries to reach equilibrium with the outdoor environment. The slope of the interior temperature change is much more shallow than that of the exterior temperature changes. The high thermal mass of the building is thought to effectively dampen the magnitude of interior temperature swings from the changes taking place in the outdoor environment.

2.3 Huntsville rammed earth residence

The final building is a two-storey residence located in Huntsville, Ontario. A typical wall section at the base of the wall to foundation connection is shown (Fig. 7). The rammed earth encloses a footprint of 178.5 m^2 (1,921 ft^2) with interior heated area of 285.7 m^2 (3,076 ft^2) and volume of approximately 1,151 m^3 (40,654 ft^3). Similar to the Castleton residence the rammed earth walls are 450 mm thick with 150 mm of polyisocyanurate insulation centrally located. Thermal mass within the envelope derives from the rammed earth walls (100,000 kg), two exposed and polished concrete slabs (70,000 kg), and a two-storey masonry stove (9,000 kg). Solar insolation is gained through 24.2 m^2 (261 ft^2, 8.5% of floor area) of south facing triple paned glazing.

Historically in Huntsville, there are 4,384 heating degree days (18°C) for Sept-April inclusive. Using the Passive House analysis method (Passive House 2014), the yearly heating energy requirement for this house volume is 39 kWhr/m^2. A similar home built to Ontario Building Code standards (OBC 2012) would result in a yearly heating requirement of about 130 kWhr/m^2.

All heat is delivered via a radiant masonry stove, although electric radiant in-slab heating was installed to meet local building code requirements. As the home is heated exclusively with wood it is difficult to accurately quantify heating energy requirements. The firewood supply is a mix of hard and soft maple, iron wood, yellow and white birch, black cherry, and small amounts of tamarack and bass wood. Wood consumed for the 2012–2013 and 2013–2014 winters was 16.31 and 23.55 m^3 respectively. Based on wood mix, moisture content, and the heater efficiency, the estimated 2012–2013 and 2013–2014 heating energy used was 35.8 kWh/m^2 and 51.6 kWh/m^2 respectively.

3 SUMMARY AND CLOSURE

Earth-based construction is a relatively new phenomenon in cold climates like those experienced in Canada. However, there is a growing body of evidence that shows that this construction technique can be successfully implemented in these locations. Both cast in-situ and compressed earth blocks have been successfully used to construct buildings up to two storeys. A key feature in a cold climate is the need for insulation. Both straw bales and more conventional polyisocyanurate insulation have been used. The performance of these buildings to date in terms of energy use has been promising, with evidence for at least one residence of heating energy requirements that are about 30% of those for a similar home of conventional construction.

REFERENCES

OBC (2012) *Ontario Building Code*, (Ministry of Municipal Affairs and Housing, Markham, 2012).
Passive House (2014) *Passive House Planning Tool*, (Passive House Institute, http://passiv.de/en/04_phpp/04_phpp.htm).

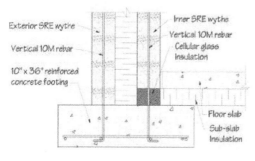

Figure 7. Typical wall cross-section for Huntsville rammed earth residence.

Rammed Earth Construction – Ciancio & Beckett (Eds)
© 2015 Taylor & Francis Group, London, ISBN 978-1-138-02770-1

Dynamic behavior of scaled Cement Stabilized Rammed Earth building models

K.S. Nanjunda Rao, M. Anitha & B.V. Venkatarama Reddy
Department of Civil Engineering, Indian Institute of Science, Bangalore, India

ABSTRACT: This paper presents the results of an experimental investigation on a half scale Cement Stabilized Rammed Earth (CSRE) building model having a plan size of 3 m × 1.7 m and height 1.5 m subjected to base motion to understand its dynamic behavior. The base motion was provided through a shock table test facility which was designed and developed as a simple and cost effective alternative to the conventional shake table test facility. Strength, elastic and damping properties of CSRE have been determined experimentally. Linear dynamic analysis using Finite Element (FE) technique has been performed for six building models having various earthquake resistant features. The response measured during the testing of the one type of building model on shock table has been compared with the computed dynamic response obtained from FE analysis. The failure patterns of the building model tested have also been presented.

1 INTRODUCTION

Rammed earth construction is receiving renewed attention by researchers across the globe due to its low carbon emission, economic viability, better aesthetics, thermal performance and availability of materials locally. There are many examples of successful application of rammed earth (both stabilized and unstabilized) for wall construction in buildings and can be found in Australia, USA, Europe, Asia and many more countries across the globe (Verma & Mehra 1950, Easton 1982, Houben & Guillaud 2003, Hall 2002, Walker et al. 2005, Jaquin 2008, Jaquin & Augarde 2011). Estimation of embodied energy in CSRE has been examined by Venkatarama Reddy & Prasanna Kumar (2010). Hall et al. (2012) have recently edited a book on modern earth buildings which discusses in a comprehensive manner aspect of materials, mechanical properties, durability issues, construction and applications. Issues of quality control and recommendations for assessment of compressive strength of CSRE have been examined by Ciancio & Gibbings (2012). Ciancio & Beckett (2013) have examined the social, financial and environmental sustainability of rammed earth.

The seismic performance of buildings made with earth are found to be far from satisfactory due its poor ability to resist cyclic action particularly the unstabilized construction. It is well known that such buildings are the most vulnerable during earthquakes due to brittleness of the material, large mass and initial stiffness, severe degradation of strength and stiffness under cyclic loading, large variability in mechanical properties of the material and poor quality of construction. The presence of openings in the walls for doors and windows further reduces their lateral load resisting capacity. Several earthen buildings have been damaged during earthquakes in the last decade like El Salvador earthquake in 2001, the Bam, Iran earthquake in 2003, the Kashmir earthquake in 2005, the Pisco, Peru earthquake in 2007, the Maule, Chile earthquake in 2010 and Van, Turkey earthquake in 2011. Several earthquake resistant features are suggested to improve the performance of such buildings during seismic events. Provision of reinforced concrete horizontal bands at different levels and integral connection of roofing system to the walls are some simple techniques which can significantly reduce their vulnerability (IS 4326: 1993). Various solutions for improving the seismic performance of earthen buildings by reinforcing them with different materials like cane, bamboo, cabuya rope, wire mesh and polymer mesh has been suggested by several investigators (Blondet & Aguilar 2007, Dowling et al. 2005, Torrealva & Acero 2005, Bartolome et al. 2008, Gomes et al. 2011).

The dynamic behaviour of adobe masonry buildings is quite different from rammed earth buildings due to differences in their modes of failure during earthquakes. To the authors knowledge the only reported study on the dynamic behaviour of rammed earth buildings is by Bui et al. (2011). Hence, the current study is undertaken to understand the response of half scale CSRE building model subjected to base motion. A linear dynamic analysis using Finite Element (FE) technique has also been performed for six building models having various earthquake resistant features.

2 EXPERIMENTAL PROGRAMME

2.1 *Details of CSRE building model*

Cement stabilized rammed earth building model of outer dimensions 3.1 m × 1.8 m × 1.5 m (length × breadth × height) with RC lintels (0.1 m width, 0.075 m depth with a bearing of 0.1 m on each side) only above the door and window openings having a wall thickness of 0.1 m was constructed on the shock table platform by compacting the processed soil in progressive layers within a temporary wooden formwork. It was constructed in three stages each of 0.5 m height. Each stage was constructed in five layers; with each layer having a thickness of 0.1 m after compaction. Figure 1 shows the plan view of the building model. Door (D) and Window (W) opening was provided on the Cross Wall (CW1) and two window openings was provided on the Cross Wall (CW2). One window opening was provided on each of the Shear Walls (SW1 & SW2). The width of door & window openings was 0.5 m and the height of door opening was 1 m and that of window opening was 0.5 m. There was no roof slab. The building model tested is designated as BM1.

2.2 *Shock table test facility*

The base motion to the CSRE building model was provided with the aid of shock table test facility, which was designed, developed and fabricated as a simple and cost effective alternative to the conventional shake table test facility. The schematic view of the shock table is shown in figure 2. It consists of a rigid steel platform of size 2.5 m × 3.5 m supported on four wheels with ability to move in one direction only. It has a pendulum of length 1.8 m and mass of 600 kg. Through pendulum hits, the platform can be set into motion. A reaction

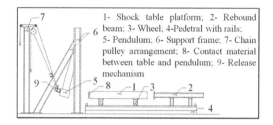

Figure 2. Schematic view of shock table facility.

steel beam is installed on the opposite side of the pendulum to provide reverse motion to the platform. The distance between the edge of the table platform and the front tip of the reaction beam can be varied. A chain-pulley arrangement is used to swing the pendulum up to a maximum of 40 degree to one side and can be suddenly released using a scissor mechanism. The characteristics of the table platform motion can be varied by changing (i) swing angle of the pendulum, (ii) mass of the pendulum, (iii) distance between the reaction beam and the edge of the table platform and (iv) contact material between the table and pendulum. It is well known that ground motions produced by an earthquake are very complicated and it can be characterized by three parameters in terms of the damage potential of an event. The parameters are (i) amplitude (ii) frequency content and (iii) duration of motion. The above parameters have been computed for the provided shock table platform motion in the present study and the same have been compared with the parameters of a few earthquake ground motions and presented later in the results and discussion section.

2.3 *Materials employed and construction of model*

Locally available soil passing through 4.75 mm sieve having sand, silt and clay fractions of 50.3%, 18.1% & 31.6% respectively was used. In order to maintain 15% clay content, the soil was reconstituted by mixing with natural river sand in the proportion of 1:1 by mass. Predominant clay mineral in the soil was kaolinite. The liquid limit and plasticity index of the reconstituted soil mix were 27% and 17.5% respectively. The soil was stabilized using 8% Ordinary Portland cement (53 grade). The maximum Proctor density and OMC for the reconstituted soil are 19.47 kN/m³ and 10.68% respectively. In the construction of the building model each layer was carefully compacted to achieve a uniform dry density of 18 kN/m³. The base of the building model was anchored to the shock table platform through a steel channel to achieve fixity at the base. The model was cured with

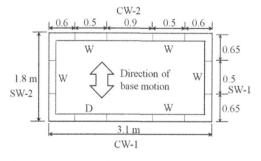

Figure 1. Plan view of CSRE building model (all dimensions in meters).

Figure 3. View of building model with instrumentation ready for testing.

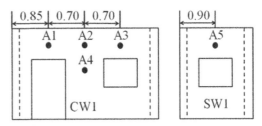

Figure 4. View showing location of accelerometers on cross and shear-walls of the building model (all dimensions in meters).

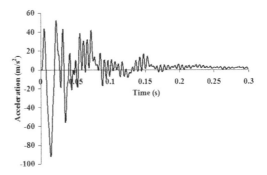

Figure 5. Time-history of shock table platform motion for 25 degree angle release of pendulum.

wet burlap for 28 days and allowed to air dry at ambient conditions before testing. Figure 3 shows a view of the building model with instrumentation before testing. Along with the construction of the model, prism and beam specimens were cast using same mix proportion to evaluate strength, elastic and damping properties of CSRE. Prisms of size 150 mm × 150 mm × 300 mm were employed to determine the compressive strength and modulus of elasticity of CSRE in dry, saturated and partially saturated conditions. CSRE beam specimens of length/height 570 mm, breadth 180 mm and depth 100 mm were employed for estimating modulus of rupture and damping. The modulus of rupture and damping were determined using four point bending test and cantilever free vibration test respectively. The specimens were tested for the cases of flexural stress parallel and perpendicular to the compacted layers.

2.4 Testing of the building model

Four numbers of piezoelectric accelerometers were fixed on each of the Cross Walls (CW1 & CW2) and one accelerometer was fixed on each of the Shear Walls (SW1 & SW2). Figure 4 shows the typical locations of accelerometers on the CW1 & SW1. One accelerometer was fixed to the platform of the shock table to measure the time history of motion. All the accelerometers had a sensitivity of 500 mV/g. The accelerometers were connected to a data acquisition system and DASYLab version 6.0 software was employed for digitizing and analyzing the data and storing it on personal computer. Free vibration test was conducted to determine the natural frequency and damping of the building model. For this purpose the building model was excited by impacting the platform of the table by swinging the pendulum by 3 degree and releasing

it. The direction of base motion was parallel to the 1.8 m side of the building model. The time history of the responses from all the accelerometers fixed to the walls of the building model were acquired and analyzed. After the free vibration test, the model was subjected to base motion by impacting the platform of the table by swinging the pendulum by an angle of 25 degree (with respect to the vertical position of the pendulum) and releasing it. Laminated ply-wood sheet of 19 mm thickness was used as contact material between pendulum and the platform of the table. The time history of the table platform motion for 25 degree angle of release of the pendulum and reaction beam at distance of 353 mm from the edge of the platform is shown in figure 5. The time history of the responses from all the accelerometers fixed to the walls of the building model were again acquired and analyzed.

3 LINEAR DYNAMIC ANALYSIS OF BUILDING MODELS USING FE TECHNIQUE

The dynamic analysis (time history method) was carried out on six building models having various earthquake resistant features. The designation of the models analyzed and the earthquake resistant features of each of them are listed below;

1. Model without roof
   a. RC lintels only above door & window openings, BM1 (same as the one tested)
   b. Continuous RC band at lintel level, BM2
   c. Continuous RC band at lintel and sill level, BM3.
2. Model with rigid roof
   a. RC lintels only above door & window openings, BM4
   b. Continuous RC band at lintel level, BM5
   c. Continuous RC band at lintel and sill level, BM6.

The buildings were modeled by 4 noded shell elemen with six degrees of freedom per node using commercially available FE software (NISA version 17). The time history of the table platform measured during the testing of the building model (25 degree angle of release of the pendulum) was used as the base motion in the FE analysis for all the building models. The fundamental natural frequency of the model BM1 obtained from FE analysis is compared with the experimentally estimated value.

4 RESULTS AND DISCUSSION

4.1 Shock table platform motion

Table 1 gives a comparison of parameters like Peak Acceleration (PA), Peak Velocity (PV), Significant Duration (SD), Housner Intensity (HI) and Arias Intensity (AI) of the shock table motion (for 25 degree angle of release of the pendulum) with Chamoli and Kobe earthquakes ground motion. It can be seen that PA and AI of shock table is very high compared with that of the earthquakes, whereas PV and HI of the shock table is quite close to that of the earthquakes. SD of the shock table is very small compared to that of the earthquakes considered. The above parameters are computed in the prototype domain. Figure 6 shows the comparison of response spectrum in the prototype domain for shock table motion with 5 degree and

25 degree angle of release of pendulum and the maximum considered earthquake (MCE) for zone 5 of IS 1893 (Part 1): 2002. It can be inferred that for short period structures having a period less than 0.3 seconds shock table test is severe enough for checking the vulnerability of structures during a big earthquake.

4.2 Strength, elastic and damping properties of CSRE

Table 2 gives various properties of CSRE. The modulus of rupture of CSRE was 1.2 MPa and 0.5 MPa for tension parallel and perpendicular to compacting layers respectively. The estimated value of damping ratio in the present study for CSRE is found to be 2.4% and it matches quite well with the value of damping for rammed earth reported by Bui et al. (2011).

4.3 Natural frequency of the building models

Table 3 gives a comparison of the fundamental natural frequency of the building model BM1 obtained from the experiment and finite element analysis. The numerically predicted value compares

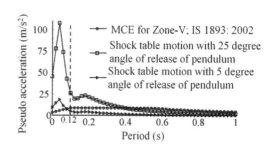

Figure 6. Response spectrum for shock table motion and MCE for zone V of IS 1893 (Part 1): 2002.

Table 2. Strength, elastic and damping properties of cement stabilized rammed earth.

Property	Dry condition	Saturated condition	Partially saturated (air dry) condition
Moisture content	1.2%	12.66%	2.92%
Compressive strength (MPa)	8.6	2.5	5.4
Initial tangent modulus (MPa)	3950	2640	3000
Strain at peak stress	0.0053	0.0015	0.0025

Table 1. Parameters of base motion.

Base motion	PA (m/s²)	PV (m/s)	SD (s)	HI (m)	AI (m/s)
Shock table platform motion	45.4	0.47	0.16	1.24	6.68
Chamoli earthquake (India) 1999	3.66	0.42	8.98	1.33	0.8
Kobe earthquake (Japan) 1995	3.38	0.27	12.86	1.42	1.68

reasonable well with measured value. The table also provides the fundamental natural frequency obtained through FE analysis for other building models having various earthquake resistant features.

4.4 *Acceleration response of the building models*

Figure 7 gives the comparison of experimentally measured acceleration response at locations A1, A2 and A3 on the cross-wall of the building model BM1 with numerically predicted values. The FE predicted acceleration response is found to have a reasonably good match with the measured response. Figure 8 shows the FE predicted acceleration response at location A4 for the various building models considered in the present study.

5 SUMMARY AND CONCLUDING REMARKS

Dynamic behavior of the half scale CSRE building model has been examined through shock table studies. The study has revealed the effectiveness of shock table test in evaluating earthquake resistance of the building model. Damping ratio of CSRE determined in this study confirms its poor energy dissipation capacity. Linear FE analysis of the building models is found to be satisfactory.

Table 3. Comparison of fundamental natural frequency of building models.

Model designation	BM1 (experiment)	FE analysis					
		BM1	BM2	BM3	BM4	BM5	BM6
Frequency (Hz)	17.1	18.8	20.5	21.2	39.9	40.1	42.3

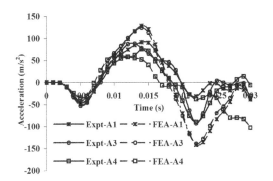

Figure 7. Comparison of experimentally measured acceleration response at locations A1, A2 and A3 with FE predicted values.

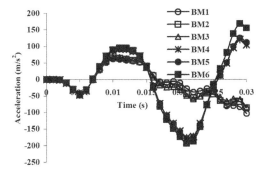

Figure 8. FE predicted acceleration response at location A4 of all the building models considered in the study.

REFERENCES

Bartolome, A.S., Quiun, D. & Zegarra, L. 2008. Performance of reinforced adobe houses in Pisco, Peru earthquake. *Proc. 14th World conf on earthquake engineering.*, Beijing, 12–17 October 2008, China.

Blondet, M. & Aguilar, R. 2007. Seismic protection of earthen buildings. *Proc. Intern Sym on earthen structures.*, Bangalore, 22–24 August 2007. India.

Bui, Q.B., Hans, S., Morel, J.C. & Do, A.P. 2011. First exploratory study on dynamic characteristics of rammed earth buildings. *Engineering Structures* 33:3690–3695.

Ciancio, D. & Beckett, C. 2013. Rammed earth: an overview of a sustainable construction material. *Proc. third intern conf. on sustainable construction materials and technologies.*, Kyoto, 18–21 August 2013. Japan.

Ciancio, D. & Gibbings, J. 2012. Experimental investigation on the compressive strength of cored and molded cement-stabilized rammed earth samples. *Construction and Building Materials* 28: 294–304.

Dowling, D., Samali, B. & Li, J. 2005. An improved means of reinforcing adobe walls-external vertical reinforcement. *Proc. Intern sem on architecture, construction and conservation of earth buildings in seismic areas.*, Lima, 16–19 May 2005. Peru.

Easton, D. 1982. *The rammed earth experience*. Wilseyville California USA:Blue Mountain Press.

Gomes, M.I., Lopes, M. & de Brito, J. 2011, Seismic resistance of earth construction in Portugal. *Engineering Structures* 33:932–941.

Hall, M. 2002. Rammed earth: Traditional methods, modern techniques, sustainable future. *Building Engineer:* 22–24.

Hall, M.R., Lindsay, R. & Krayenhoff, M. (eds). 2012. *Modern earth buildings: Materials, engineering, construction and application.* Sawston, Cambridge, UK: Woodhead Publishing.

Houben, H. & Guillaud, H. 2003. *Earth construction: A comprehensive guide.* London: Intermediate Technology Publication.

IS 1893 (Part 1): 2002. Criteria for earthquake resistant design of structures. New Delhi. India: Bureau of Indian Standards.

IS 4326: 1993. Earthquake resistant design and construction of buildings-code of practice. New Delhi. India: Bureau of Indian Standards.

Jaquin, P. 2008. Study of historic rammed earth structures in Spain and India. *The Structural Engineer* 86:26–32.

Jaquin, P. & Augarde, C. 2011. *Earth Building: History, Science and Conservation*. Berkshire: IHS BRE Press.

Torrealva, D. & Acero, J. 2005. Reinforcing adobe buildings with compatible mesh: The final solution against the seismic vulnerability?. *Proc. Intern sem on architecture, construction and conservation of earth buildings in seismic areas. Lima*, 16–19 May 2005, Peru.

Venkatarama Reddy, B.V. & Prasanna Kumar, P. 2010. Embodied energy in cement stabilized rammed earth walls. *Energy and Buildings* 42:380–385.

Verma, P.L. & Mehra, S.R. 1950. Use of soil-cement in house construction in the Punjab. *Indian Concrete Journal* 24:91–96.

Walker, P., Keable, R., Martin, J. & Maniatidis, V. 2005. *Rammed earth: Design and construction guidelines*. UK. IHS BRE Press.

Rammed Earth Construction – Ciancio & Beckett (Eds)
© 2015 Taylor & Francis Group, London, ISBN 978-1-138-02770-1

Rammed Earth construction for the Colorado Front Range region

R. Pyatt
Program in Environmental Design, University of Colorado, Boulder, Colorado, USA

C. O'Hara
Studio NYL Structural Engineers, Boulder, Colorado, USA

R. Hu
School of Architecture, Xi'an University of Science and Technology, Xi'an, Shaanxi, China

ABSTRACT: Rammed Earth (RE) has been used as a traditional building method for thousands of years throughout the hot dry desert climate of the American Southwest. The feasibility of rammed earth as a contemporary construction method in cold climates is illustrated through the presentation of a single case study with a focus on the Colorado Front Range region. Rammed earth construction methods, formwork challenges, design details for high wind and seismic loads, optimized thermal performance strategies, and the durability of soil mixtures for freeze/thaw cycles are explored to establish the viability of rammed earth as a sustainable and contemporary building method for the colder climate zones along the Colorado Front Range region and throughout the west.

Keywords: Rammed Earth; stabilized rammed earth; insulated rammed earth; post-tensioned rammed earth; contemporary rammed earth construction

1 INTRODUCTION

Buildings are responsible for more than 40 percent of global energy used, and as much as one third of global greenhouse gas emissions, [UNSP-SBCI, 2009]. As the cost of energy rises, and rapid urbanization increases, it has become imperative to take advantage of low cost passive strategies to reduce the energy demand of our buildings, improve the resiliency of our local communities and contribute to a positive energy future.

1.1 Casa sanitas: A Rammed Earth prototype

Casa Sanitas (a *healthy home*) is part of a rammed earth research project at the Program in Environmental Design at the University of Colorado Boulder, and is the first rammed earth house to be constructed in the city of Boulder, Colorado.

Conceived as a sustainable alternative to the conventional wood frame houses found across the Colorado Front Range. The Casa Sanitas prototype is designed to be a 'positive energy home' (positive energy homes produce more energy over the course of a year than they use) and combines cost effective passive design strategies that include the high thermal mass of RE walls, natural ventilation and passive solar orientation, with

PhotoVoltaic Panels (PV), solar thermal hot water and a small ground source heat pump for radiant heating and cooling.

A key objective of the project is to develop a comprehensive rammed earth case study to help inform a set of "best practices" for rammed earth construction in the colder climate of Colorado, specifically around optimized thermal performance, as well as to establish an "applied research" laboratory to educate the community in the design and construction of sustainable, affordable, and regionally appropriate housing.

Phase two of the research project will include installing a custom data acquisition system to monitor the house over several years to collect data on the energy performance of the RE prototype over multiple seasons, as well as to improve energy modeling capabilities for high thermal mass RE homes in the Colorado Front Range region.

Additional research supported by the National Natural Science Foundation of China [project number: 51378410] will include exploring the feasibility of contemporary post-tensioned rammed earth wall systems in rural China through a research collaboration between the School of Architecture at Xi'an University of Science & Technology and the Program in Environmental Design at the University of Colorado Boulder.

1.2 Location & climate

The Colorado Front Range is mainly a transition zone found at the base of the Rocky Mountains from the eastern plains. Temperatures fluctuate from hot and dry to cold and wet. Precipitation can be minimal or severe (NOAA, 2014).

The project is located on a standard suburban lot in Boulder Colorado along the central Front Range at 5900 feet in elevation. This zone is characterized by a cold winter and hot dry summer, and with a large temperature fluctuation during the day and night. The high thermal mass of rammed earth walls is particularly beneficial for buildings in a hot dry climate with large diurnal temperature variations.

2 RE WALL CONSTRUCTION

2.1 Soil selection

The soil for the wall is comprised of a local engineered fill (road base) material blended with local crushed granite (crusher fines) to achieve an even particle size distribution appropriate for a RE wall (Fig. 5). The soil was stabalized using 8% portland cement by weight which make the wall more durable in weathering and freeze/thaw cycles. The RE walls were built two years ago and have been exposed to severe weather conditions common to Boulder and remained unaffected.

2.2 Structural design

The structural design of the RE walls is based on the International Residential Code 2006 edition, the amendments made by the State of New Mexico, and the ACI 318 (American Concrete Institute). The system has been designed as a post-tensioned wall using unreinforced low compressive strength concrete as the baseline.

For design purposes the RE walls were designed with an allowable compressive strength (Fc') of 500 psi. Confirming test cylinders during construction exceeded 2,200 psi. The wall is utilized for both vertical and lateral loads imposed on the system. Given the mass of the wall and the relatively light vertical loads applied to it the gravity loads were not the primary design concern, but rather the out of plane seismic loads and the in plane seismic and wind loads applied to the wall. As the wall is unreinforced, it was assumed to have negligible tensile capacity. The system is post-tensioned to ensure that the rammed earth does not exhibit any net tension.

The post-tensioning system consists of sleeved threaded rods in each of the two wythes of rammed earth not more than 30 feet apart. The rods are embedded in the foundation concrete with a double

nut to increase area of pressure on the concrete. Above to capping bond beam a 4" × 4" plate and nut are utilized to clamp the system. Following compaction and curing of the capping bond beam the nut is torqued to hand tight. Then an extension piece is added to the wrench to provide additional torque so that 1.5 more revolutions are possible beyond hand tight. The thread pitch on the rod is 20 threads per inch essentially elongating the rod by 1/15th of an inch. This results in a force of approximately 10,000 lbs in the rod resulting in a pre-stress in the rammed earth of a net 5-psi. The maximum tension in this seismic zone for out of plane loads on the wall is less than 3 psi. In order to use of the RE walls as shear walls the system is designed based on an assumed control joint pattern of 12 ft on center. Establishing a wall system consisting of (5) 12-foot long shear walls for calculation purposes. The significant length of shear wall available and the loads imparted on a single story building resulted in only 2 psi of tensile load resulting from wind load and only 1 psi for seismic loads.

2.3 Frost Protected Shallow Foundation

Given the significant thickness of the wall system, the use of a conventional foundation system is not economically feasible. Typical foundations in this region would need to be a minimum of 36" below grade around the perimeter of the building to avoid frost heave. This would result in approximately 0.25 cubic yards per linear foot of concrete around the perimeter of the building. The use of the Frost Protected Shallow Foundation (FPSF) reduced the amount of concrete required to 0.08 cubic yards per linear foot and contributed to the affordability of the system.

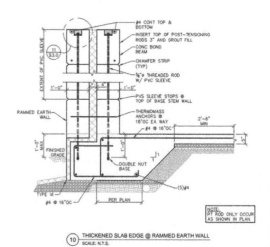

Figure 1. Detail of frost protected shallow foundation.

The system has been designed based on SEI/ASCE 32-01 American Society of Civil Engineers: Design and Construction of Frost Protected Shallow Foundations. Essentially the system creates a barrier to frost propagation through the soil by insulating the perimeter grade around the building and tying that directly to the buildings perimeter wall insulation to prevent any thermal bridge through the system that could make the building susceptible to frost heave. Horizontal insulation around the perimeter of the house extends 16" away from the wall using R10 XPS rigid foam board insulation, and extends 24" at the foundation corners. In order to link the line of perimeter insulation with the insulation encapsulated within the double wythe rammed earth walls, a horizontal line of insulation transitioned across the concrete base wall at approximately the slab level (Fig. 2). The outer wythe of concrete at this location is anchored to the inner wythe using fiberglass connectors to prevent sliding of this element in a seismic event and to provide positive anchorage.

2.4 RE wall construction method & formwork

Standard modular concrete forms were used in the construction of the RE walls, however, accommodating the double wythe wall with interior insulation using standard forming systems posed a significant construction sequencing challenge. It was determined that compacting the wall in stages was more efficient than trying to accommodate the insulation in the center at one time.

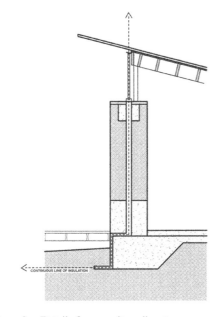

Figure 2. Detail of composite wall system.

Figure 3. Photograph of Casa Sanitas under construction.

The outer 12" wythe was combined with the 4" insulation board along the inside of the form using a standard 16" form tie. The soil was mixed and delivered to the forms with a standard bobcat and compacted using pneumatic tampers in 8" lifts. Once the outer wythe of the double wall was complete the form ties were cut and the inside formwork was adjusted to accept the interior soil mix using custom field welded form tie extensions. This allowed the outside formwork to stay in place but increased the time required in the field to create custom form ties. Current improvements to the overall construction system include research specifically addressing the need for specialized formwork for insulated rammed earth walls to allow for seamless adjustment of the interior formwork contributed to overall affordability of the system.

3 STRATEGY FOR ENERGY EFFICIENCY & INDOOR COMFORT

3.1 Thermal properties of the RE wall

Thermal mass is an essential strategy to achieving indoor comfort particularly in a hot arid region with high diurnal temperature variations. Rammed earth walls have high thermal mass. The rammed earth can contain or absorb more heat than concrete does even when it is less dense (Soebarto, 2009). The internal rammed earth wall can provide a long thermal time lag to stabilize the indoor temperature. A rammed earth wall can also stabilize the indoor humidity to keep the indoor environment comfortable.

However, a rammed earth wall without insulation has a low R-value of 0.4/inch (CSIRO, 2000). Materials with high thermal resistance R-values can reduce heat flux under steady state conditions to reduce the energy demand. With the introduction of interstitial insulation a higher R value is achieved in a RE wall and the thermal performance of the wall as a whole is much improved (Hall and Allinson, 2008). A composite RE wall has high thermal mass on both sides of a core of insulation and combined with a high R value insulated roof and high quality door and windows, the building envelope can ensure the house will be extremely energy efficient, keeping the house cool in the summer and warm in the winter.

To take advantage of the thermal mass and address the poor thermal performance of the wall, four inches of rigid polystyrene foam (XPS) insulation was added at the center of the RE assembly (Fig. 2). Thermal tests show that this type of composite wall system incorporating XPS insulation in the center of a double wyth wall system performs better in colder climates than an un-insulated RE wall (Hall and Allinson, 2008).

Aligning the line of insulation in the center of wall with that of the triple-glazed windows above continues the separation across the wall-to-window transition eliminating any potential thermal bridge (Fig. 2). Thermal analyses of Casa Sanitas, using THERM software, showed an ideal separation of interior and exterior temperatures across the wall section.

Moreover, using three years of real weather data for Boulder, Colorado, the hygrothermal model of the composite wall system indicated that the introduction of the R 20 rigid insulating layer results in all temperature variation occurring in the outer wythe of the wall, while the interior wythe indicated only a 10-degree variation throughout the entire year and most surprisingly only a degree or two within any individual day (Fig. 4).

4 CONCLUSION

A significant concern with any composite wall system is moisture build-up in the interior or exterior wythe, which could cause indoor air quality issues and structural damage. Hygrothermal analyses of the RE composite wall using WUFI software noted that the drying potential for the system designed for the Casa Sanitas prototype, is quite high given the Colorado Front Range climate. The most recent 3-year sample weather data for Boulder showed that any condensation in the wall assembly was restricted to the outer layers, with the dew-point temperature never being surpassed except in the exterior RE wythe or within the 4 inches of XPS insulation (Fig. 4). Also, due to the wall assembly's drying potential for the local climate, there was no moisture accumulation in any of the layers.

Given these results, a composite wall system that includes four inches of (R-20) rigid insulation in the center of a double wythe RE wall assembly is recommended as a "best practice" for climate responsive contemporary RE construction in Boulder, Colorado. Additionally the high thermal mass of the interior wyth of RE is particularly appropriate as part of a low cost passive strategy to reduce the energy demand of the prototype home and contribute to a positive energy future.

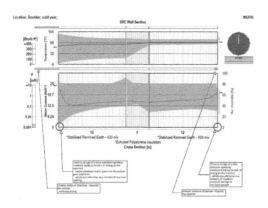

Figure 4. WUFI analysis of composite wall system.

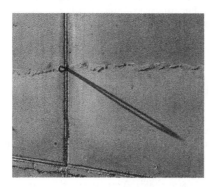

Figure 5. Photograph of Casa Sanitas wall.

REFERENCES

CSIRO Media Release (2000) Ref 2000/110. Available online: http:/www.dab.uts.

Hall, M.A. and Allinson, D. 2008. Assessing the moisture-content-dependent parameters of stablized earth materials using the cyclic-reponse admittance method. Energy, Build. 2008. 40, 2044–2051.

NOAA. 2014. Earth Systems Research Laboratory. Available online: http://www.esrl.noaa.gov/psd/boulder/.

SBCI, *United Nations Environment Programme, 2009*. A report on buildings and climate change. Available online: http://www.unep.org/sbci/pdfs/sbci-bccsummary.pdf.

Soebarto, V. Analysis of indoor perfomance of houses using rammed earth walls, Eleventh International IBPSA conference, 2009. Available online: http://www.ibpsa.org/proceedings/BS2009/BS09_1530_1537.pdf.

Windstorm, Bly & Schmidt, Arno. A report of contemporary rammed earth construction and research in North America. SUSTAINABLILITY 2013, 5, 400–416; doi: 10.3390/.

Rammed Earth Construction – Ciancio & Beckett (Eds)
© 2015 Taylor & Francis Group, London, ISBN 978-1-138-02770-1

Analytical model for predicting the stress-strain behaviour of Cement Stabilised Rammed Earth

L. Raju & B.V. Venkatarama Reddy
Indian Institute of Science, Bangalore, India

ABSTRACT: The strength and elastic properties of Cement Stabilised Rammed Earth (CSRE) greatly depend upon the cement content and density for a given soil composition and grading. The paper deals with investigations on stress-strain characteristics of CSRE considering two different densities and cement contents. An analytical model developed closely predicts the stress-strain response of CSRE.

1 INTRODUCTION

Rammed earth is a monolithic construction and is constructed by compacting processed soil in progressive layers in a temporary formwork. Both load bearing and non-load bearing walls can be built using rammed earth. Two types of rammed earth constructions can be recognised: stabilised rammed earth and un-stabilised rammed earth. Stabilised rammed earth contains inorganic additives such as cement or lime. Cement has been used for rammed earth walls since the last five to six decades. Examples of cement stabilised rammed earth for buildings can be seen in Australia, USA, Europe, Asia and many other countries (Verma & Mehra 1950, Easton 1982, Houben & Guillaud 2003, Matthew Hall, 2002, Walker et al. 2005).

It is possible to get guidelines and specifications from the literature on the soil and stabilisers for the rammed earth wall constructions. The past studies on stabilised rammed earth indicate a range of strength values and recommend use of sandy soils with cement content in the range of 6–15%. Apart from strength and durability characteristics it is essential to ascertain stress-strain relationships and elastic properties of rammed earth for: (1) assessing the strength and stability of rammed earth structures under concentric and eccentric loads and (2) predicting the behaviour of rammed earth structures/elements under different types of loading conditions.

There are limited studies on stress-strain characteristics and elastic properties of CSRE. Venkatarama Reddy and Prasanna Kumar (2009 & 2011) examined the stress-strain relationships for CSRE considering different densities (1600–2000 kg/m³) and cement contents (5–12%). These studies revealed that modulus of CSRE is sensitive to moisture content, cement content and

density, and the ultimate failure strains for CSRE in dry condition stretch up to 2%. Bahar et al. (2004) report initial tangent modulus value of 2.51 GPa for 10% cement rammed earth in dry condition.

There are hardly any studies on developing analytical models for predicting the stress-strain response for CSRE. Hence, the present investigation is focused on generating stress-strain relationships for CSRE and developing an analytical model to predict the stress-strain response.

2 METHODOLOGY

Rammed earth cylindrical specimens were prepared using two cement contents (7 and 10%) and two different dry densities (1650 and 1800 kg/m³). After 28 days curing the specimens were air dried and then oven dried at low temperature (50°C). The dried specimens were tested in dry as well as in saturated condition in a displacement controlled testing machine while recording the strains. This data was used to plot stress-strain relationships. Analytical model was developed to predict the stress-strain behaviour of CSRE.

3 MATERIALS USED IN THE EXPERIMENTS

Ordinary Portland Cement (OPC) conforming to IS 12269 (1987) was used for casting the CSRE specimens. 28 day compressive strength of OPC tested following the procedure outlined in IS 4031 (1988) was 69.2 MPa. The initial and final setting time for the cement was 148 and 312 minutes respectively. A reconstituted local red soil was used for casting the rammed earth specimens. Comprehensive investigations of Venkatarama Reddy and Prasanna Kumar (2011) on cement stabilised rammed earth revealed that the optimum clay content in the soil yielding maximum strength is about 15%. Therefore, natural red soil was reconstituted by mixing the soil and sand in the proportion of 1:1 (by mass). Reconstituted soil contains sand, silt and clay size fractions of 72.6%, 11.6% and 15.8% respectively. Predominant clay mineral in the soil was kaolinite. The liquid limit and plasticity index of the reconstituted soil mix are 26.9% and 17.5 respectively. Standard Proctor OMC and maximum dry density for the soil were 10.28% and 1992 kg/m³ respectively.

4 CASTING CYLINDRICAL SPECIMEN AND TESTING PROCEDURE

Rammed earth cylindrical specimens of size 150 mm diameter and 300 mm height were used for determining compressive strength and stress-strain relationships of CSRE. The following procedure was adopted for preparing the rammed earth specimen.

a. The crushed and dried soil mix was blended with cement. Requisite quantity of water (OMC) was sprayed onto the soil-cement mixture and mixed thoroughly, and it was ensured that the cement and water were distributed uniformly in the mixture.
b. The partially saturated mix was poured into a metal cylindrical mould and compacted in three layers of 100 mm thickness. A flat headed rectangular shape (3 kg) and a flat headed rounded corner shape (1.5 kg) rammers were used for compaction. The mass of the material in each layer was controlled such that the final designated dry density of the cylindrical specimen was achieved.
c. The specimen was removed from the mould after 24 hours of casting and kept for curing under wet burlap. After 28 days of curing the specimens were allowed to dry in air inside the laboratory for two weeks. The air dried specimens were then oven dried at 50°C to attain constant weight and then were used in the experiments.

Figure 1 shows the experimental set-up showing the extensometer positioned in the middle of the specimen to record strains over a gauge length of 100 mm. The specimens were tested in a displacement controlled testing machine at the displacement of rate of 3 microns per second. The tests were conducted in both oven dry and saturated state. The oven dried specimens were soaked in water for 48 hours prior to testing in order to saturate the specimen. After the test the moisture content of the failed specimens were determined.

Figure 1. Experimental set-up.

5 RESULTS AND DISCUSSION

5.1 Strength and stress-strain characteristics of CSRE

Figures 2 and 3 show the stress strain relationships for CSRE using 7 and 10% cement respectively. These curves represent the mean of four specimens tested in each case and give stress-strain cures for dry and saturated cases as well as for the two dry densities (1650 and 1800 kg/m³). Modulus and dry density relationships for the CSRE are shown in Figure 4. Figure 5 shows a combined plot for strength and modulus representing both dry and saturated cases. The strength and stress-strain characteristics of CSRE are given in Table 1. The following points emerge from the results shown in these Figures and the Table.

1. The stress-strain relationships are linear initially followed by non-linear portion until peak stress for all the cases. Post peak relationships show drooping curves. Post peak response in dry condition shows considerable deformation indicating larger strains at failure (0.75–1.0%). The strain at peak stress (ε_u) is more in dry condition than in saturated condition irrespective of cement content and density. ε_u values are more for the lower density (1650 kg/m³) specimen.
2. The modulus of CSRE is sensitive to density, water content and cement content. The Initial Tangent Modulus (ITM) of CSRE with 1800 kg/m³ dry density is about 2.8 times the modulus of CSRE with 1650 kg/m³ dry density.

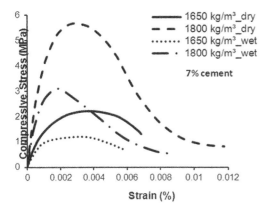

Figure 2. Stress-strain relationships for CSRE (7% cement).

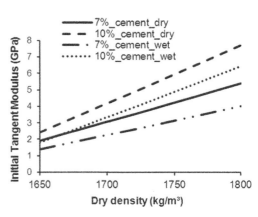

Figure 4. Modulus versus dry density.

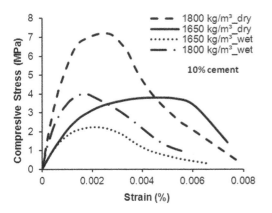

Figure 3. Stress-strain relationships for CSRE (10% cement).

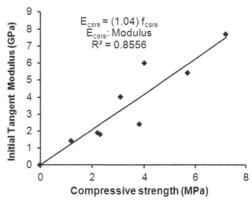

Figure 5. Modulus versus compressive strength.

Table 1. Stress-strain characteristics of CSRE.

γ_d (kg/m³)	7% cement						10% cement					
	Dry			Wet			Dry			Wet		
	ITM (GPa)	ε_u	f_{csre} (MPa)	ITM (GPa)	ε_u	f_{csre} (MPa)	ITM (GPa)	ε_u	f_{csre} (MPa)	ITM (GPa)	ε_u	f_{csre} (MPa)
1650	1.9	0.0039	2.21	1.4	0.0030	1.19	2.4	0.0047	3.82	1.8	0.0024	2.33
1800	5.4	0.0031	5.71	4.0	0.0018	3.10	7.7	0.0025	7.22	6.0	0.0017	4.00

γ_d—Dry density; ITM—Initial Tangent Modulus; ε_u—Strain at peak stress; f_{csre}—Cylinder compressive strength.

For CSRE having 1800 kg/m³ dry density and 7% cement content the ITM is 4.0 GPa and 5.4 GPa for wet and dry cases respectively. For 10% cement the corresponding values are 6.4 and 7.7 GPa.

3. There is a linear relationship between modulus and cylinder compressive strength (Figure 5) with a correlation coefficient of 0.86. The initial tangent modulus (in GPa) is 1.04 times the cylinder compressive strength (in MPa).

5.2 Analytical model for prediction of stress-strain behavior of CSRE

The stress-strain response of the CSRE can be predicted using the following analytical model.

$$\sigma_{csre} = f_{csre}\left(\frac{\varepsilon_c}{\varepsilon_{csre}}\right)^k \frac{n}{(n-1)+\left(\dfrac{\varepsilon_c}{\varepsilon_{csre}}\right)^{nk}} \quad (1)$$

where "n" and "k" are curve fitting factors;
σ_{csre} = Compressive stress on CSRE;
f_{csre} = Compressive strength of CSRE cylinder;
ε_{csre} = longitudinal strain at peak stress;
ε_c = strain at corresponding stress level f_c.
$n = 1.45 + \frac{f_{csre}}{9.98}$ for dry specimens and
$k = 1$, when $\frac{\varepsilon_c}{\varepsilon_{csre}} \leq 1$ for saturated specimens.
$k > 1$, when $\frac{\varepsilon_c}{\varepsilon_{csre}} > 1$ and $k > 1$, when $\frac{\varepsilon_c}{\varepsilon_{csre}} > 1$.

This model predicts the stress-strain response of CSRE in both dry and saturated state. This model has different expressions for the curve fitting factors when compared to the expression proposed by Thorenfeldt et al. (1987) for high strength concrete. The curve fitting factor "n" is a function of compressive strength of CSRE cylinder. Compressive strength in turn is a function of dry density and cement content of CSRE cylinder. Based on the experimental stress-strain curves the expressions for "n" in terms of strength were derived.

Figures 6 and 7 show the stress-strain relationships predicted using the proposed model and the experimental stress-strain relationships for typical cases. The analytical model proposed predicts the stress-strain response for CSRE very closely.

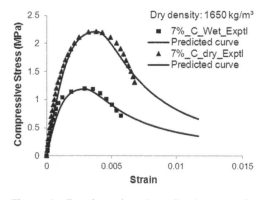

Figure 6. Experimental and predicted curves for 1650 kg/m³.

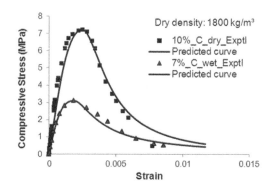

Figure 7. Experimental and predicted curves for 1800 kg/m³.

5.3 Suction pressure, strength and stiffness

The stress-strain relationships presented in the investigation are for dry and saturated conditions. Influence of suction pressure on the strength and stiffness of CSRE is absent at these two extreme moisture conditions of the CSRE specimen. Suction arises because of capillary meniscus formation between particles. When degree of saturation = 0 (i.e. dry condition) no water is available for meniscus formation and hence suction pressure = 0. Likewise for degree of saturation = 100% all voids are saturated and no meniscus formation possible and hence suction pressure = 0.

6 CONCLUDING REMARKS

Stress-strain relationships for CSRE were established considering two different densities and cement contents in both dry and saturated conditions. The proposed analytical model closely predicts the strain-strain response for CSRE. The modulus and strength relationship is of the form: Initial tangent modulus (in GPa) = (1.04) × (cylinder compressive strength in MPa). The investigations can be extended to encompass wide range of densities and cement contents.

REFERENCES

Bahar, R., Benazzoug, M. & Kenai, B. (2004). Performance of compacted cement-stabilised soil. *Cement & Concrete Composites* 26: 811–820.

Easton, David. (1st Ed.) (1982). *The rammed earth experience*. Wilseyville, CA: Blue Mountain Press.

Hall, Matthew. (2002). *Rammed earth: Traditional methods, modern techniques, sustainable future*, Building Engineer: 22–24.

Houben, H. & Guillaud, H. (2004). *Earth construction–A comprehensive guide*, London: Intermediate Technology Publications.

IS 12269 (1987), *Specification for 53 grade ordinary Portland cement.* Bureau of Indian standards, New Delhi, India.

IS 4031 (part 7) & (part 5) 1988, *Methods of physical tests for hydraulic cement.* Bureau of Indian standards, New Delhi, India.

Walker, P., Keable, R., Martin, J. & Maniatidis, V. (2005). *Rammed earth design and construction guidelines*, BRE Bookshop UK.

Thorenfeldt, E., Tomaszewicz, A. & Jensen, J.J. (1987). Mechanical Properties of High Strength Concrete and Application in Design. *Proc. of Symposium on Utilization of High Strength Concrete*, Tapir, Throndheim, 149–159.

Venkatarama Reddy, B.V. & Prasanna Kumar, P., (2009). Compressive strength and elastic properties of stabilised rammed earth and masonry, *Masonry International*, 22(2): 39–46.

Venkatarama Reddy, B.V. & Prasanna Kumar, P. (2011). Cement stabilised rammed earth—Part B:Compressive strength and elastic properties, *Materials and Structures*, 44(3): 695–707.

Verma, P.L. & Mehra, S.R. (1950). Use of soil-cement in house construction in the Punjab. *Indian Concrete Journal*, 24: 91–96.

Rammed Earth Construction – Ciancio & Beckett (Eds)
© *2015 Taylor & Francis Group, London, ISBN 978-1-138-02770-1*

Advanced prefabricated rammed earth structures—mechanical, building physical and environmental properties

J. Ruzicka, F. Havlik, Jan Richter & Kamil Stanek
Department of Building Structures, Faculty of Civil Engineering, CTU in Prague, Prague, Czech Republic

ABSTRACT: Long construction times, labor intensive technology, high risks of technological faults, limited on-season time and volume changes during the curing period are some of the disadvantages of using rammed earth in modern structures. Prefabrication can eliminate those disadvantages and can also bring environmental benefits in decreasing of negative impact of site works on the environment. Rammed earth itself can effectively contribute to the thermal and RH quality of internal microclimate. The paper summarizes latest results of experimental research of mechanical, building physical and environmental properties of precast rammed earth.

1 INTRODUCTION

1.1 Rammed earth in modern environmental effective structures

Environmental advantages of rammed earth as a construction material are recyclability and low energy demands of manufacturing which leads to lower embodied energy and emissions.

The hygroscopic properties of clay allow effectively absorb and release moisture and makes earth an ideal material to moderate the indoor microclimate from the point of view of RH. Jokl (1991) states that long-term stabilization of relative humidity influences positively occupant's health.

Rammed earth as a raw natural material also helps to keep higher level of ions in the air and can effectively shield electromagnetic smog against penetration into building when applied in sufficiently thick layer as Minke (2001) states. Positive experience of application in hospitals, psychiatric facility, chapels was observed as generally spoken presence of natural material positively stimulate human mind.

One of latest examples of using rammed earth to keep stabile microclimate is the new Herb Centre for Ricola Company in Laufen by Basel, Switzerland designed by Herzog & de Meuron (Figs. 1 and 2). The building was finished in 2013. The façade consists of prefabricated rammed earth panels. In the stocking part earthen panels keep the stabile microclimate for storing of herbs without any other ventilation system which decreases the energy demand in operation phase of the building, makes it energy efficient and decreases negative environmental impact. The range of use is great as the building dimensions are approx. 50×30 m with almost 10 m height.

Figures 1 & 2. Ricola Herb Centre, Laufen by Basel, Switzerland, Herzog & de Meuron Architects, construction phase, July 2012. Production of prefabricated rammed earth elements and application on building site. Development, production and construction of rammed earth structures were made by Martin Rauch, LEHM TON ERDE.

1.2 Development of advanced prefabricated rammed earth structures at CTU in Prague

The main objective of the long term project was to verify possibilities of using prefabricated rammed earth panels for load bearing structures.

Ruzicka & Havlik (2012) describe the technological process of manufacturing and application of prefabricated wall panels using rammed earth core and wooden frame which was verified within the construction of a low-energy family house in Pilsen, Czech Republic (2008). The building was designed as a timber structure with a load bearing timber columns. The diaphragm interior wall creating heat accumulator was designed as a pre-formed from wall panels of the size 950 × 650 × 200 mm.

Positive experience from this first pilot project led to the development of the second generation of prefabricated rammed earth structures. Load bearing elements for vertical structures without wooden frame and any reinforcement were designed and tested within the research project.

Preliminary tests for optimizing the mixture were provided on large sets of small scale test samples. The optimized mixture was used for manufacturing and testing of single prefabricated elements and finally the load bearing capacity of the story-high wall consisting of prefabricated elements was tested.

Four main topics regarding development of prefabricated rammed earth elements have been recognized as crucial for the development: (i) shrinkage behavior of clay from the point of view of final structural and aesthetical quality of rammed earth; (ii) mechanical properties and size effect; (iii) building physical properties from the point of view of sorption potential and accumulation potential of earthen structures; (iv) environmental properties.

Stabile material source was used for all performed tests during the whole project period. The earth is used by Claygar company in the Czech Republic for clay plaster and unburned bricks production. The basic mixture consists of Hydromicas and Kaolinite with minor addition of Montmorillonite. Pneumatic rammer with electric air pump was used for manufacturing rammed earth samples.

2 SHRINKAGE PROPERTIES

2.1 Determination of shrinkage properties of earth mixtures

One of the main problems of using earth in building structures is shrinkage during the curing time and volume changes under the influence of air humidity.

About 53 sets of samples differed in water content, granulometry (sand content), mechanical

and chemical stabilization have been tested to find an optimal mixture to eliminate shrinkage and to optimize other mechanical properties (compressive strength, bending tension strength, static modulus of elasticity in compression). Special tests in climatic chamber have been carried out to determine volume changes under the RH influence. The tests have been provided according to ČSN EN 12617-4. A raw material was dried out to control the water content and the admixture amount. Test samples of the size 40 × 40 × 160 mm with special cogs enabling precise length measuring at a dilatometer have been placed at both ends of each specimen. After 24 hours the samples were extracted from covered forms and placed in a room with monitored temperature and RH. The results were measured on 3 samples during a period of 28 days. Values of shrinkage are transformed to general unit mm/m. Mechanical properties were tested on the same samples. The determination of shrinkage properties of earth was carried out in three levels: (i) shrinkage properties under the water content influence; (ii) under the influence of mechanical, physical and chemical stabilization; (iii) rheological changes under the influence of RH changes. Detailed results are published in Ruzicka & Havlik (2012).

2.2 Results

From the beginning of the project it was obvious that the mixture for prefabricated wall panels should be without any chemical stabilization like cement, lime or other additives to keep the environmental quality of the structure. From this point of view only water content and sand stabilization or addition of natural based reinforcement was accepted.

The tests show that most of the shrinkage occurs in the first 7 days; afterwards the progression is much slower. The final shrinkage was determined after 28 days of drying into equilibristic state. It is obvious from the results that water content control is crucial to avoid or minimize shrinkage. Water content in the levels 8, 10, 12, 15, 20% (by weight) was tested. If water content is 8% the workability of the mixture is difficult due to low cohesion, stronger compaction is necessary. Shrinkage is on a very low level, also mechanical properties are decreasing. If water content exceeded 15% the mixture starts to be muddy, fabrication is also difficult. Shrinkage is on the highest level and mechanical properties are also poor (Tab. 1). From the point of view of shrinkage the optimal water contents seems to be in the range from 10% to 12% even if the differences of final shrinkage are almost 30%! It is also obvious that the amount of water from 8% to 20% can reduce

Table 1. Shrinkage and mechanical properties under the influence of water and sand content. Introduction to the samples marking: C—pure clay loam; W—water; S—sand.

Sample	Mixture	Density [kg/m³]	Shrinkage in 28 days [mm/m]	Compress. strength [MPa]
Water content:				
C_W8	8% water	2139	17.74	8.94
C_W10	10% water	2189	22.55	10.63
C_W12	12% water	2163	32.48	8.77
C_W15	15% water	1945	51.11	5.81
C_W20	20% water	1946	67.23	6.10
Sand stabilization (water content 10%):				
C_S10/W10	10% sand	2152	18.82	7.88
C_S20/W10	20% sand	2164	13.19	6.84
C_S30/W10	30% sand	2164	7.03	6.22
C_S40/W10	40% sand	2141	4.52	4.75

or increase the shrinkage properties more than 3.5 times!

Mechanical stabilization by sand influences shrinkage properties by changing of granulometry (decreasing amount of clay parts) and makes the mixture drier. For determination of influence of this kind of stabilization water content was kept constant (10%). Sand of the fraction 0–4 mm was used and added in the values of 0, 10, 20, 30, 40%. This reduces shrinkage from 22.55 mm/m (0% of sand) up to 4.52 mm/m (40% of sand) as shows in Table 1.

Adsorption ability and volume sensitivity under the influence of RH changes was tested in climate chamber. The crack development in joints is important for prefabricated structures to keep aesthetical quality and also air tightness in case of low energy and passive buildings. The test was carried out in three steps at constant temperature of 20°C. The first step took 72 hours at 33% RH and the specimens have been equilibrated into steady state; the second step took 48 hours at 75% RH and the third 48 hours at 33% RH. Weight and volume changes were monitored in every step. The swelling for the mixture C_W12 was +1.2 mm/m when RH increased from 33% to 75% in first 72 hours and shrinkage was −0.8 mm/m in 48 hours when RH decreased again to 33%. The results for C_S30/W12 were +0.9 mm/m buckling in 72 hours and −0.4 mm/m shrinkage in next 48 hours. The positive influence of sand stabilization was observed.

All above mentioned tests were provided also for other kinds of stabilization as adding of mechanical particles as sawdust, cellulose, Poraver, chopped straw, chemical stabilization by cement, lime were tested. Ruzicka & Havlik (2012) state detailed description of the results of shrinkage properties.

3 MECHANICAL PROPERTIES AND SIZE EFFECT PROBLEM

3.1 Size factor

The size effect is well known problem linked to testing of mechanical properties of structural materials. Testing of earthen materials and preparing test samples is compared to concrete, steel etc. very labor intensive and time demanding process. Finding the reliable correlation between the size of test samples and final compressive strength could help in real design and assessment of earth structures. The aim of this part of the project was to find size factor for the mixture C_S30/W11 used for the panels. Compressive strength at 4 kinds of samples was compared: cubes 40 mm (remains of beams 40 × 40 × 160); cubes 100 mm, 150 mm and 200 mm (each set of 3 samples). The results for calibration curve are accompanied by compressive strength of cylinders 150 × 300 mm and single panels of the size 1000 × 600 × 200 mm. The technology of ramming, mixture recipe and boundary condition were identical in all cases.

Almost linear dependence of sizes 40–100–150 mm is obvious in Fig. 4. The formula given by ČSN EN 12390-3 for size factor of concrete shows that compressive strength of cube 200 mm is 95% of strength of cube 150 mm. This relation is also included in Fig 3. Conversion factors for each set and different approaches are shown in Table 2 where reference size of cube is 150 mm. As it can be seen in Fig. 3, the trend for rammed

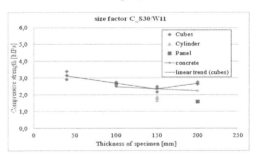

Figure 3. Trends of influence of specimen size to compressive strength, mixture C_S30/W11.

Figure 4. Assembling of the testing wall and final collapse.

Table 2. Compressive strength conversion factor.

Thickness [mm]	40	100	150	200	Cylinder 150
C_S30/W11					
Average	1.34	1.12	1.00	1.14	0.75
Trends	1.34	1.15	1.00	1.13	–
Concrete (standards)	–	1.05	1.00	0.95	0.8

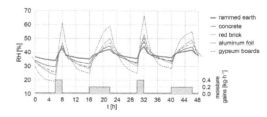

Figure 5. Calculated course of relative humidity in the test room over a period of 2 days.

earth is steeper comparing to concrete. The results show that the cube 100 mm is 12% stronger in compressive strength than cube 150 mm. But also cube 200 mm is 12% stronger than 150 mm. This is probably given by effect of optimal thickness to layer depth in bigger specimen.

3.2 Load bearing capacity of the wall

The testing wall of the size $1.8 \times 3.0 \times 0.2$ m consisted of 10 prefabricated panels. The wall was assembled in 2 days after the drying period of 6 month and load bearing capacity was determined. Slip tongue connection with clay mortal was used on vertical joints, clay mortal of the thickness 3 mm on horizontal joints. The concentrated load was uniformly distributed through stiff steel beam. The wall was subjected to axial compression. The wall failure occurs in upper horizontal joint and was initiated by lateral deflection. Peak failure load was 524.1 kN, (load bearing capacity 1.6 N/mm²), final vertical deformation 1.2% (37 mm).

4 BUILDING PHYSICAL PROPERTIES

Although natural clay has excellent sorption properties, in practical use clay materials contain certain amount of sand primarily to minimize the shrinkage. Typical clay plasters contain between 70–75% of sand which decrease rapidly sorption ability. Such a high amount of sand means that the sorption properties of clay plaster are at the same level as traditional lime-cement plasters as Minke (2006) states. On the other hand the sand fraction in unburned bricks and rammed earth structures usually does not exceed 30%, which makes these structures more effective in terms of interior moisture moderation.

Within the development of prefabricated rammed earth panels at CTU, the sorption curves of three rammed earth mixtures containing 0%, 10% and 30% of sand were measured and presented by Richter J. et al (2014). The results show that if added 10% of sand (mixture C_S10/W10) the sorption curve drops by 12.5% compared to the mixture C_W10 (without sand). If added

Table 3. Environmental parameters for prefabricated rammed earth panels.

Parameters	Prefab ram. earth	Clay at mine	Clay plast. at plant	Concr. normal at plant	Brick at plant
PEI [MJ/kg]	0.1963	0.0439	0.4819	0.5749	2.5737
GWP [kgCO$_{2,ekv}$/kg]	0.2391	0.0029	0.0191	0.1099	0.2386
AP [gSO$_{2,ekv}$/kg]	0.0217	0.0224	0.0716	0.1849	0.5456
Vol. weight [kg/m³]	2000	2000	1815	2380	600

30% of sand (mixture C_S30/W10) the sorption curve drops by 28% compared to C_W/10.

The determined sorption curve for the mixture C_S30/W10 was used for further numerical analysis and was compared with other structural materials. Richter et al (2014) describes the methodology and comments the results. It shows that rammed earth can effectively moderate the indoor air humidity variations; calculated daily variation of RH in testing room: rammed earth 9%, concrete layer 15%, gypsum boards 19%, red bricks 23% (Fig. 5).

5 ENVIRONMENTAL PROPERTIES

Environmental impact connected with the construction of the buildings is valuated mostly by: embodied CO_2 [gCO$_2$/kg] as a global environmental load (GWP), embodied SO_2 [gSO$_2$/kg] as a local environmental load (AP), embodied energy (PEI) [MJ/kg] or total weight of the constructions [t]. Calculation of embodied emissions and energy for prefabricated panels compared to other clay materials and concrete (Tab. 3) is based on data for "clay at mine" and "sand at mine" sourced from Ecoinvent database (2012) and emissions produced during the manufacturing process were added. This calculation considers only "cradle to gate" part which includes (i) material transportation from mines to the lab, (ii) drying and grinding, (iii) mixing and ramming.

However the unit parameters of earth and clay show much higher environmental quality compared to other building materials (bricks, concrete) the real impact is lower because in real structures the volume weight and the total weight of the structure has to be taken into account. The real potential for the improvement is about 20%.

6 CONCLUSIONS

This project represents one of possible approaches to sustainable building using earth structures as a modern technology. The results, mathematical simulation and also the practical examples show positive influence of rammed earth to RH of internal microclimate and technological potential and environmental properties of precast rammed earth.

ACKNOWLEDGEMENTS

This outcome has been financially supported by internal grants of CTU SGS14/113/OHK1/2T/11 "Analysis and optimization of properties of natural building materials effecting quality of interior microclimate of buildings".

REFERENCES

CSN EN 12617-4 Products and systems for the protection and repair of concrete structures—Test methods—Part 4: Determination of shrinkage and expansion (in Czech).

ČSN EN 12390-3. Testing hardened concrete—Part 3: Compressive strength of test specimens (in Czech).

Jokl, M. 1991, Indoor Environmental Quality. 2. ed., Prague: CTU in Prague (in Czech).

Minke, G. 2001. Das Neue Lehmbauhandbuch: Baustoffe, Konstruktionen, Lehmbauarchitektur, Staufen bei Freiburg, Ökobuch.

Minke, G. 2006. Building with earth, Design and architecture of a Sustainable Architecture. Basel: Birkhäuser.

Richter, J., Stanek, K., Ruzicka, J. & Havlik, F. 2014. Sorption properties of rammed-earth mixtures, Civil Engineering Journal, Vol. 5–6. (in Czech).

Richter, J., Stanek, K., Ruzicka, J. & Havlik, F. 2014. Stabilization of indoor humidity by rammed earth. Advanced Material Research, Vol. 1000 Ecology and New Building Materials and Products, pp. 342–345.

Ruzicka, J. & Havlik, F. 2012. Advanced Prefabricated Rammed Earth Elements for Vertical and Horizontal Structures. In *Proceedings of LEHM 2012—Tagungsbeiträge der 6. Internationalen Fachtagung für Lehmbau.* Weimar. Dachverband Lehm e.V., pp. 134–145.

www.ecoinvent.ch

Rammed Earth Construction – Ciancio & Beckett (Eds)
© *2015 Taylor & Francis Group, London, ISBN 978-1-138-02770-1*

The role of clay and sand in the mechanics of Soil-Based Construction Materials

J.C. Smith & C.E. Augarde
School of Engineering and Computing Sciences, Durham University, UK

ABSTRACT: Clay is a key component of most Soil Based Construction Materials (SBCMs) (a term we use in preference to Rammed Earth to cover all insitu and unit-based methods), and is often referred to as a "binder" for these materials. To make the most of SBCMs there may be occasions when non-ideal clays (such as those with expansive properties) are used, to avoid transportation of better alternatives to site. This paper investigates, through the use of X-Ray Computed Tomography (XRCT) and unconfined compression testing, the effects of adding expansive clay to a soil mixture. Changes in, what we term, the macrostructure of the material are discussed and are linked to observed drying, shrinkage and compressive strength properties. The effect of changing the proportions of clay and sand within the mix is also investigated to better understand their roles. Two main experiments are described. Both are performed on small triaxial samples, 38 mm diameter & 76 mm long, of two mix compositions using different clay mixes (a pure Kaolin clay; and a 80% Kaolin 20% Bentonite clay). The first experiment determined the development of unconfined compressive strength in the samples as they dried and the effect different clay types had on this development. The second experiment used XRCT to scan the samples and determine the change in macrostructure as samples dried. The results obtained from the experiments demonstrate minimal changes in unconfined compressive strengths of SBCMs when small amounts of expansive clays are included and the effect on the development of macrostructure of the material.

1 INTRODUCTION

Soil Based Construction Materials (SBCMs), a term used to cover all insitu and unit-based methods of compacted earth, can be considered, when in its un-stabilized form, as a highly unsaturated soil in which the main source of its strength is suction (Jaquin et al. 2008) and therefore the role of clay as a 'binder' in this material is crucial. The use of expansive clays is often avoided within SBCMs as the material is known to crack during drying (Walker et al. 2005) adversely affecting the final strength of the structure. However, clay found on site can often contain a small amount of expansive material and it is environmentally sensible not to import material if possible and hence make use of an expansive clay. This paper investigates the effect of a small amount of expansive clay within an unstabilized SBCM, particularly focussing on the unconfined compressive strength (referred to as just compressive strength from here) and the changes in Void Size Distributions (VSDs), during drying.

Two main experiments, performed on small triaxial samples of compacted SBCM mixes, are discussed in detail in this paper. The first experiment investigates the development of compressive strength within the samples as they dry, particularly considering the affect the addition of a small amount of expansive clay has on the suction developed when dry and the final compressive strength. The second experiment uses X-Ray Computed Tomography (XRCT) to determine the evolution of the samples' VSD during drying and whether cracking can be observed when expansive clay is used. Conclusions are then drawn about the suitability of small amounts of expansive clay in an unstabilized SBCM mix.

XRCT is a non-destructive 3D imaging technique capable of imaging and analysing internal structures within solid samples, to a resolution of less than one micron. A typical laboratory XRCT machine contains three elements: the X-ray source as a conventional X-ray tube; a sample stage, which rotates the sample to enable a series of X-ray images to be obtained at incremental angular positions; and a detector in the form of a scintillator screen followed by a CCD camera (Helliwell et al. 2013). For a detailed description of the essentials of XRCT the reader is referred to Ketcham & Carlson (2001) for an insight into various issues, such as image artefacts and edge detection, that arise when using XRCT for quantitative analysis of materials. The non-destructive nature of XRCT and its ability to scan large samples to a macropore resolution has made it an ideal technique for the investigation of SBCMs. In recent years XRCT has been used to investigate the mesopore size distribution of cement stabilised soils (Hall et al. 2013), structural changes of unsaturated soil under loading (Beckett et al. 2013) and very recently Smith & Augarde (2014b) investigated the VSD of a single Rammed Earth mix.

2 EXPERIMENTATION

2.1 Materials and manufacture

Two different clay:sand mixes were investigated in this study, both of which combined sand (<2 mm) and clay at a ratio of 1:2 by dry mass. Gravel was not included in any of the mixes as it is necessary to sieve out the dry soil fraction greater than 2.36 mm to prevent these larger particles interfering with compaction and obscuring the XRCT images (Tarantino 2009). One clay fraction contained only Speswhite Clay (a pure kaolin clay supplied by IMERYS Performance Minerals), referred to hereafter as K100, whilst the second contained 80% speswhite and 20% Wyoming Sodium Bentonite (a highly expansive clay supplied by RS Minerals Ltd), referred to hereafter as K80. Dry density (ρ_d) and optimum water content (w_{opt}) values were obtained via the British Standard Vibrating Hammer Test, suggested by Smith & Augarde (2014a) to produce the closest match to compaction regimes used during construction, and these values, plus further geotechnical parameters, i.e. Liquid Limit (LL), Plastic Limit (PL) and Linear Shrinkage (LS), are given in Table 1.

Five stages of drying were investigated and are denoted here using the dryness factor (df), an indication of the how close the sample is to its dry state given the current atmospheric conditions. The df value of each sample is given as a value between zero and one, whereby a sample of $df = 1$ is in its dry state whereby no further water loss is measured and a sample of $df = 0$ is at w_{opt}. Filter paper tests were performed, using the calibration equation proposed by Hamblin (1981), for each stages of drying enabling total suction values to be measured across the range of df values.

Sixteen samples were manufactured for each mix at the maximum ρ_d, three per drying stage and one XRCT sample. Each conformed to standard triaxial dimensions (38 mm diameter, 76 mm height) and was formed in two equal layers, each statically compacted at w_{opt}. They were then left to dry, in a temperature monitored room at 22°C ± 2°C, until they reached specific values of df and were then sealed using plastic caps and a latex sheath. The sealed samples were left to equilibrate for a minimum of 24 hours.

2.2 XRCT scanning & analysis

All XRCT images were obtained using the Zeiss Versa XRM410 XRCT scanner installed at Durham University School of Engineering and Computing Sciences. Scans were performed on one sample for each mix, repeated at three stages of drying ($df = 0,0.5,1$), and involved a full sample scan and a higher resolution Region Of Interest (ROI) scan at the centre of the sample to obtain data for both layers and the interface. The XRCT key scan parameters applied can be found in Table 2.

The analysis process was performed using Avizo Fire software and an automated analysis procedure written for analysing the samples. Threshold values for the voids and solids were found autonomously by identifying the peaks within the sample histograms and watershed segmentations were used to identify individual voids. The analysis produced a data file in which each void was assigned a unique ID and information including volume, maximum length and minimum width was collected for all the voids identified. From this it was possible to produce VSD plots for all of the samples scanned, noting that the pixel size in Table 2 represents the smallest void detectable.

2.3 Mechanical testing

Unconfined compression, constant water content tests were performed on the 30 samples not XRCT scanned using a Lloyd LR5K Plus Testing Machine with a load cell rated at 5 kN ± 0.5%. Each sample was tested to failure, observed by evidence of a load peak, after which compression was immediately stopped and the load released. The tests were

Table 1. Geotechnical parameters for K80 & K100.

Mix	ρ_d (kg/m³)	w_{opt} (%)	LL (%)	PL (%)	LS (%)
K80	1.56	11.4	52	17	11.9
K100	1.60	12.3	36	19	6.8

Table 2. The XRCT parameters used for scanning.

Scan type	Pixel size (μm)	Field of view (mm)	Scan time (hr)
Full	20.2	42.4 × 42.4	7
ROI	2.1	4.2 × 4.2	19

performed at rate of 0.1 mm/minute to ensure all tests took approximately 30 minutes. Dry density values were calculated for the samples at the point of testing, and all were within 1.2% of the target dry density values.

3 RESULTS & DISCUSSION

3.1 Mechanical behaviour

Figure 1 shows the development of compressive strength of both the K80 and K100 material as the samples dry. The large spread of results is systematic of the complex nature of the structure of SBCMs, even in laboratory conditions, and the inherent variability introduced when compacting a soil mixture, however the well-established trend of increase of compressive strength as the water content decreases and the samples dry is visible. Presented in this format however it is hard to draw many firm conclusions, although it may be possible to say that the K80 material has a higher compressive strength than the K100 dried samples at water contents below 8% (i.e. ignoring samples close to compaction water content) and so the presence of small amounts of expansive clay within a SBCM *may* be beneficial to the compressive strength properties of a SBCM.

However, when the average compressive strength is plotted against *df*, and it is possible to see how the materials develop in strength relative to their end state (the water content at which the materials reach equilibrium with the atmospheric conditions), the results become a little clearer. It can be seen that the addition of the small amount of expansive clay, in the K80 samples, has had no considerable effect on the compressive strength of the material as it dries and, perhaps most importantly, when both materials reach their dry state the compressive strength of the K80 samples is

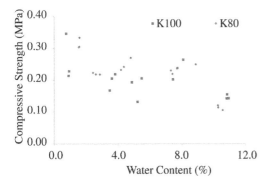

Figure 1. The compressive strength of all 30 samples wrt. Water content.

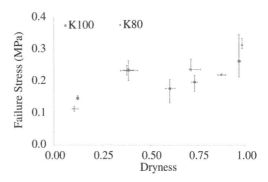

Figure 2. The average compressive strength plotted against average *df* at all five stages of drying.

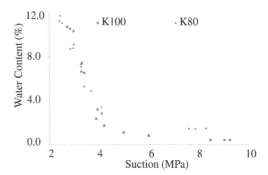

Figure 3. The development of total suction during drying.

higher. From an industry perspective this finding tentatively suggests that there is no need to avoid using small amounts of expansive clay in SBCMs, when considering the compressive strength.

3.2 Development of suction

The compressive strength results can be further examined using the filter paper data which provide a measure of the suction, the main source of strength in unstabilized SBCMs. Figure 3 shows that at highest water contents there is no clear difference between the suctions developed in the two materials and therefore provides an explanation as to why there is no noticeable difference in the compressive strength values at the higher water contents. From the trend of the results it is also apparent that at lower water content K80 exhibits higher suctions than K100 and this would explain the larger compressive strengths observed for the K80 samples.

However, Figure 3 also shows that the K100 mix ultimately develops higher suction values which would suggest a higher compressive strength, a result not observed in the mechanical testing.

To explain this it is once again necessary to plot the results against *df*, as shown in Figure 4. Once this has been done it is quite evident that there is no difference between the two mixtures in terms of the development of suction as the K80 and K100 materials reach equilibrium at their dry state and since compressive strength tests were performed at the same stages of drying for both mixes, rather than water contents, the similarity in mechanical behaviour can be better understood.

These suction results further confirm the suggestion from the compressive strength results that there is no requirement to avoid using small amounts of expansive clay in SBCMs, when considering the compressive strength properties of the material.

3.3 *Changes in internal structure*

Figure 5 shows the VSD obtained for five of the six full sample XRCT scans performed on the two materials. The K100 *df* = 0 result is not shown due to an error in the scanning process preventing

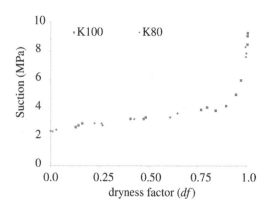

Figure 4. The development of total suction wrt dryness.

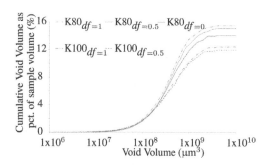

Figure 5. The VSD of K80 & K100 samples obtained using the full sample XRCT scans.

comparison of the results with the others shown. The detailed samples are currently being analysed and these results will be presented in detail at the conference. The difference in VSDs of the two mixes is a function of the different materials used as well as the compaction. Therefore, here we focus on the change in a sample's VSD during drying rather than comparing different samples.

It is clear in both materials the voids increase in size as the sample dries although the reasons for this are not clearly established. It is however crucial to note that no cracks, which would have been identified as very large voids during the XRCT scans, were observed in either samples during drying and it is suggested that the increase in void size is due to the clay matrix present in the material shrinking as the material dries and therefore causing an increase in the volume of voids within the sample. The positive Linear Shrinkage (LS) values presented in Table 1 support this hypothesis, showing both materials' fractions less than 425μm shrink upon drying, and it is believed that the presence of the sand grains in both mixes will have limited the degree to which the materials shrink thus preventing cracking. This is the subject of on going work and will be also presented in more detail at the conference.

These XRCT results, on these particular materials, suggest that cracking does not occur despite a small increase in the size of the voids during drying. Therefore it may not be necessary to avoid the use of a small amount of swelling clay in unstabilized SBCMs.

4 CONCLUSIONS

The effect of the presence of a small amount of expansive clay into unstabilized SBCMs has been investigated by considering the changes in compressive strength, the development of suction and the VSD at a range of drying stages. Firstly it has been suggested that the presence of expansive clay does not adversely affect the compressive strength of the material, particularly when considering the results with respect to their drying state—not water content. Secondly the total suction values have been shown to be higher, for a given water content, in the SBCM samples where the expansive clay is present and the lack of adverse effects on the mechanical properties further explained when the development of suction in both materials was shown to be the same for any given stage of drying. Thirdly, through the use of XRCT and determining the VSD for each material as it dries, it has been shown that cracking does not occur when a small amount of expansive clay is present however small increases in void sizes are evident in both materials as they dry. It is therefore concluded that it is not necessary to immediately discount for use in SBCM any soil where expansive clay is found, as small amounts of expansive clay, such as the ones used in this study, may not adversely effect the compressive strength of the structure or its tendency to crack.

ACKNOWLEDGEMENTS

We would like to acknowledge the assistance provided by the Durham XRCT Facility, which was funded in part by the EPSRC (grants EP/K036084/1 & EP/K024698/1).

REFERENCES

Beckett, C., M. Hall, & C. Augarde (2013). Macrostructural changes in compacted earthen construction materials under loading. *Acta Geotechnica 8*(4), 423–438.

Hall, M., S. Mooney, C. Sturrock, P. Matelloni, & S. Rigby (2013). An approach to characterisation of multi-scale pore geometry and correlation with moisture storage and transport coefficients in cement-stabilised soils. *Acta Geotech-nica 8*(1), 67–79.

Hamblin, A. (1981). Filter-paper method for routine measurement of field water potential. *Journal of Hydrology 53*, 355–360.

Helliwell, J., C. Sturrock, K. Grayling, S. Tracy, R. Flavel, I. Young, W. Whalley, & S. Mooney (2013). Applications of x-ray computed tomography for examining biophysical interactions and structural development in soil systems: a review. *European Journal of Soil Science 64*(3), 279–297.

Jaquin, P., C. Augarde, & L. Legrand (2008). Unsaturated characteristics of rammed earth. *Unsaturated soils: advances in geo-engineering: proceedings of the 1st European Conference on Unsaturated Soils*, 417–422.

Ketcham, R. & W. Carlson (2001). Acquisition, optimization and interpretation of x-ray computed tomographic imagery: applications to the geosciences. *Computers & Geo-sciences 27*(4), 381–400.

Smith, J. & C. Augarde (2014a). Optimum water content tests for earthen construction materials. *Proceedings of the ICE -Construction Materials 167*(CM2), 114–123.

Smith, J. & C. Augarde (2014b). Xrct scanning of unsaturated soil: Microstructure at different scales. *Geomechanics from Micro to Macro, CRC Press*, 1137–1142.

Tarantino, A. (2009). A water retention model for deformable soils. *Géotechnique 59* (9), 751–762.

Walker, P., R. Keable, J. Martin, & V. Maniatidis (2005). Rammed earth: design and construction guidelines. *BRE Bookshop, Watford*.

Rammed Earth Construction – Ciancio & Beckett (Eds)
© *2015 Taylor & Francis Group, London, ISBN 978-1-138-02770-1*

On the relevance of neglecting the mass vapor variation for modelling the hygrothermal behavior of rammed earth

L. Soudani, A. Fabbri, P.A. Chabriac & J.-C. Morel
LGCB-LTDS, UMR 5513, CNRS, Université de Lyon, ENTPE, France

M. Woloszyn & A.-C. Grillet
LOCIE, UMR 5271, CNRS, Polytech Annecy-Chambéry, Université de Savoie, France

ABSTRACT: Earthen materials might be a solution for energy issues being faced in construction, given their abilities to buffer moisture and improve indoor air quality while keeping the internal temperature relatively stable. However, their impact on the global energy performance of the buildings remains unclear. Our study aims at quantifying the effects of heat and mass transports, and phase changes occurring within the pores, on the hygrothermal behavior of earthen materials. To achieve this, a coupled hygrothermal model is derived. The relevancy of a certain simplifying assumption, i.e. neglecting the effects of mass vapor variation, is assessed through the analysis of the accuracy of the modeling, studied by numerical simulations on COMSOL Multiphysics®.

1 INTRODUCTION

The building sector plays a key role in greenhouse energy consumption. In fact, most of these building materials, either for insulation or for wall manufacturing, are major energy consumers during both their production and implementation (embodied energy), and their recycling is not always operational (Harris, 1999). Consequently, the development of the earth based buildings appears to be a sustainable alternative to conventional constructions (Sameh, 2013).

In addition, one of the main assets of earthen materials is their role in moisture buffering and temperature controlling, which can be related to phase change processes occurring within the pores (Liuzzi, Hall, Stefanizzi, & Casey, 2013). To integrate all those processes into a complete model, some authors (Gray, 1983; Whitaker, 1977) start from a microscopic scale and reach the macroscopic scale by averaging on a representative volume, allowing a better appraisal of the assumptions required. Others (Künzel, 1995; Luikov, 1975; Philip & De Vries, 1957) have adopted a phenomenological approach enabling them to deal with physical problems right from the macroscopic scale. This latter even gave rise to commercially developed software (Fraunhofer, n.d.),(Grünewald, 1997) which can provide reliable results on a wide range of materials and climatic loads. However, these models, which are based on simplified transport and storage functions, may exhibit some difficulties to reproduce with accuracy the hygrothermal behavior of unconventional materials like earth when they are submitted to important hygrometry and temperature variations.

The aim of this paper is to quantify the influence of the mass vapor variation in time, a particular assumption made while deriving the governing equations, in order to examine their degree of relevance in their application to the study of the hygrothermal behavior of earthen materials.

2 HYGROTHERMAL MODEL SET-UP

In this paper, the rammed earth is modelled as the superposition of a solid skeleton (S) and a porous network partially saturated by liquid water (L), assumed to be pure. The remaining porous network space is filled by a continuous gaseous phase (G), which is assumed to be an ideal mixture of perfect gases composed of dried air (A) and water vapor (V).

2.1 Liquid water-vapor equilibrium

The local equilibrium between liquid water and its vapor implies the equalities of their specific free enthalpies. Assuming that the total pressure of the gaseous phase remains constant, equal to the atmospheric pressure, this assumption allows relating the liquid pressure to the relative humidity of ambient air via the celebrated Kelvin's Law:

$$p_G - p_L = -\frac{\rho_L RT}{M_{H_2O}} \ln \varphi \qquad (1)$$

Differentiation of the Kelvin Law then leads to the two following equations:

$$dp_L = \frac{\rho_L RT}{M_{H_2O}} d\ln\varphi + \frac{\rho_L R\ln\varphi}{M_{H_2O}} dTdp_V$$

$$= p_V^{sat}(T)d\varphi + \varphi\frac{dp_V^{sat}}{dT}dT \qquad (2)$$

2.2 Water continuity equation

Assuming no air flow within the porous network, the mass conservation of water vapor reads:

$$\frac{\partial m_V}{\partial t} = -\nabla\cdot\left(\rho_V\phi_G\left(\underline{V}_V - \underline{V}_G\right)\right) + m_{\to V}^\circ \qquad (3)$$

where $m_{\to V}^\circ$ is the rate of vapor mass production due to evaporation/condensation processes. The term $\rho_V\phi_G(V_V - V_G)$ stands for the diffusive transport of water vapor within the gaseous phase, which can be evaluated through the Fick Law.

$$\frac{\partial m_V}{\partial t} = \underline{\nabla}\cdot\left(\frac{RT}{M_{H_2O}}D_e^V\nabla(p_V)\right) + m_{\to V}^\circ \qquad (4)$$

Assuming the relative velocity of the liquid in the porous media follows the generalized Darcy Law, conservation of liquid water mass can be expressed as:

$$\frac{\partial m_L}{\partial t} = \rho_L\underline{\nabla}\cdot\left(-\frac{kk_r^L}{\eta_L}\nabla p_L\right) - m_{\to V}^\circ \qquad (5)$$

Water in the material is present both as liquid and vapor. Using (2)(4)(5), and neglecting the variation of water content with temperature at constant humidity, the balance of the overall water mass reads:

$$\begin{cases} \dfrac{1}{\rho_L}\left(\rho_d\dfrac{\partial w}{\partial\varphi} + \phi_G\dfrac{M_{H_2O}p_V^{sat}(T)}{RT}\right)\dfrac{\partial\varphi}{\partial t} + \dfrac{\phi_G\rho_V^\alpha}{\rho_L}\dfrac{\partial T}{\partial t} \\ \quad = \nabla\cdot\left(\mathcal{K}^T\underline{\nabla}T + \mathcal{K}^\varphi\underline{\nabla}\varphi\right) \\ \rho_V^\alpha = \varphi\dfrac{p_V^{sat}(T)M_{H_2O}}{RT}\left(\dfrac{1}{p_V^{sat}(T)}\dfrac{dp_V^{sat}(T)}{dT} - \dfrac{1}{T}\right) \\ \mathcal{K}^T = \dfrac{kk_r^L}{\eta_L}\dfrac{\rho L^R}{M_{H_2O}}\ln\varphi + D_e^V\dfrac{\varphi M_{H_2O}}{\rho_L RT}\dfrac{dp_V^{sat}(T)}{dT} \\ \mathcal{K}^\varphi = \dfrac{kk_r^L}{\eta_L}\dfrac{\rho_L RT}{M_{H_2O}\varphi} + D_e^V\dfrac{M_{H_2O}}{\rho_L RT}p_V^{sat}(T) \end{cases}$$

$$(6)$$

2.3 Equation of heat transfer

The enthalpy balance of the wall leads to the thermal equation in its classical form for porous media with in-pore water phase change (e.g., (Fabbri, Coussy, Fen-Chong & Monteiro, 2008)):

$$\frac{\partial H}{\partial t} = \nabla\cdot\left(\lambda\underline{\nabla}T\right) - m_{\to V}^\circ L(T,\varphi) \qquad (7)$$

Where H the average enthalpy at constant pressure given by:

$$H = T\left((1-\phi)\rho_S C_{p,S} + \phi S_r\rho_L C_{p,L} + \phi(1-S_r)(\rho_A C_{p,A} + \rho_V C_{p,V})\right) \qquad (8)$$

L is the latent heat associated to the liquid/vapor phase change taking place at conditions different from the reference state. Using (1), it can be expressed as follows as a function of temperature and equilibrium relative humidity:

$$L(T,\varphi) = L_0 + (C_{p,V} - C_{p,L})(T - T_{ref}) + (\alpha_L T - 1)\frac{RT}{M_{H_2O}}\ln\varphi \qquad (9)$$

To solve the thermal equation, we need an additional relation on the evaporation/condensation rate, which can be evaluated either by equations (4) or (5). This evaluation is theoretically equivalent, in accordance with the overall mass conservation equation.

However, numerical problems can occur depending on which form is used, and the second option requires an evaluation of the term $\partial m_V/\partial t$. In (Künzel, 1995), this term is simply neglected and only the first term of the right side of eq. (4) is accounted for. It implies to consider the heat source due to phase changes as proportional to the divergence of the water vapor diffusion flux density. A further investigation is needed in order to find the best compromise between a more complicated form or a simpler one but which needs some additional assumptions. The influence of this choice on the hygrothermal coupling will be studied in the following. In the end, it can lead to two final forms for the heat transfer equation: the first one using the relation (5) and the second one relation (4).

$$\begin{cases} \dfrac{\partial H}{\partial t} - \left(L(T,\varphi)\rho_d\dfrac{\partial w}{\partial\varphi}\right)\dfrac{\partial\varphi}{\partial t} = \nabla\cdot\left(\mathcal{L}^T\underline{\nabla}T + \mathcal{L}^\varphi\underline{\nabla}\varphi\right) \\ \mathcal{L}^T = \lambda - L(T,\varphi)\rho_L\dfrac{kk_r^L}{\eta_L}\dfrac{R}{M_{H_2O}}\ln\varphi \\ \mathcal{L}^\varphi = -L(T,\varphi)\rho_L\dfrac{kk_r^L}{\eta_L}\dfrac{RT}{M_{H_2O}\varphi} \end{cases}$$

$$(10)$$

$$\begin{cases} \dfrac{\partial H}{\partial t} - = \underline{\nabla} \cdot \left(\mathcal{L}^T \underline{\nabla} T + \mathcal{L}^\varphi \underline{\nabla} \varphi \right) \\[2mm] \mathcal{L}^T = \lambda + L(T,\varphi) \dfrac{RT}{M_{H_2O}} D_e^V \varphi \dfrac{dp_v^{sat}(T)}{dT} \\[2mm] \mathcal{L}^\varphi = L(T,\varphi) \dfrac{RT}{M_{H_2O}} D_e^V p_V^{sat}(T) \end{cases}$$

$$(11)$$

Let us note that the first formulation, through the term $d\varphi$, requires the use of the sorption isotherm, whose accuracy is not robust for high relative humidity.

3 RESULTS AND DISCUSSIONS

The model predictions have been compared to experimental data for a rammed earth wall. The experimental process is described more precisely in (Chabriac, Fabbri, Morel, Laurent, & Blanc-Gonnet,

2014). This comparison gives some confidence on the ability of the model to simulate accurately the hygrothermal behavior of rammed earth walls.

3.1 Numerical evaluation of the assumption's influence

To underline the main assets of the coupled model developed in this study and thus to identify the singularity of the hygrothermal behavior of rammed earth, we need to simulate other and more complicated loads than those produced experimentally. To do so, two numerical experiments, respectively referenced as LP1 and LP2 are considered, solving equations on a 50 cm thick wall (1.5×1 m²) placed between two insulated boxes.

The loading path LP1 considers daily relative humidity sinusoidal cycles between 70% and 50% at a constant temperature of 30°C within the insulated box. The loading path LP2 considers daily temperature cycles between 0°C and 20°C at a

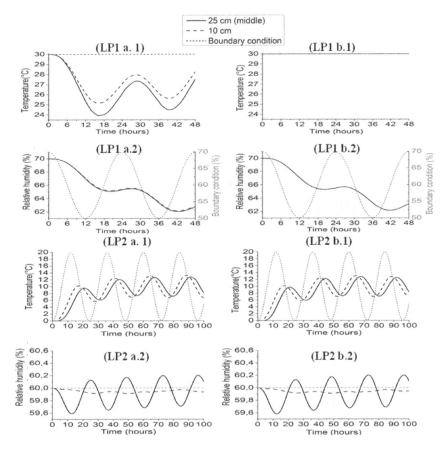

Figure 1. Simulated temperature and relative humidity distributions in both cases for LP1 and LP2, for different points in the thickness of the wall.

constant relative humidity of 60%. For each test conditions, 2 simulations are considered. The first one (referenced as "a") is based on the system of equations (6)(10) and the second one (referenced as "b") on system (6)(11). Results of LP1 and LP2 for the two kinds of simulations in temperature and relative humidity are reported in Figure 1.

No significant modifications are observable when we consider the loading path LP2. This result is not surprising as the variations in relative humidity remain slight: the evaporation process is obviously rather driven by the flow of water vapor than the mass variation of water vapor within the pores.

On the contrary, results from the LP1's conditions are much more different between the simulations. Indeed, regarding temperature distributions (LP1a.1 and LP1b.1), hygrothermal effects vanish when equation (4) with negligible vapor mass variation assumption is chosen. Actually, under this kind of solicitation, the variations in relative humidity induce a significant increase of mass of water vapor within the porous network. And this variation becomes no more negligible when compared to the flow of water vapor though the porous network. Again, this behavior may be caused by earth particularities, whose porosity is quite important towards its effective diffusion coefficient. This conclusion should not be then generalized for all construction materials.

4 CONCLUSION

A coupled model, capable of simulating the heat and mass transport, taking into consideration effects due to phase change of water inside the earthen walls, is developed. The main advantage of this model is to consider separately the kinematics of each phase (e.g., liquid water, vapor, dry air and solid matrix), in interaction with each other. It also accounts for the impact of pore water pressure on the liquid-to-vapor phase change, and hence on the resulting latent heat released or absorbed.

The model is used to assess the accuracy and impacts of a simplifying assumption made by the hygrothermal models for buildings materials. It follows that, due the singularities of the materials considered (from very low water content to near saturation, high porosity, large variation of water content during its life-time, ...), the variation of vapor mass shouldn't be neglected when strong variations of relative humidity take place in the material.

5 NOMENCLATURE

$C_{p,I}$ [J.kg^{-1}.K^{-1}]: specific heat capacity at constant pressure for component I

D_e^V [m^2.s^{-1}]: effective diffusion coefficient
I: whether liquid (l), solid (s), vapor (v)
L [J.kg^{-1}]: latent heat associated with liquid/vapor phase changes
M_I [kg]: mass of component I
$m_{\to V}^\circ$ [kg.s^{-1}]: rate of vapor mass production due to phase changes
M_{H_2O} [kg.mol^{-1}]: molar mass of water
P_I [Pa]: partial pressure of component I
p_V^{sat} [Pa]: equilibrium vapor pressure
R [J.K^{-1}.mol^{-1}]: gaz constant
S_r [-]: saturation ratio
T [K]: temperature
V_I [m.s^{-1}]: relative velocity of component I in the porous media
W [-]: mass water content
α_L [-]: thermal volume dilatation coefficient of the liquid
η_L [Pa.s]: dynamic viscosity of water
K [m^2]: intrinsic permeability of the porous medium
k_r^L [-]: relative liquid permeability
λ [W.m^{-1}.K^{-1}]: thermal conductivity
ρ_I [kg.m^{-3}]: density of component I
φ [-]: relative humidity
ϕ [-]: porosity

ACKNOWLEDGMENTS

The present work has been supported by the French Research National Agency (ANR) through the "Villes et Bâtiments DUrables" program (Project Primaterre no. ANR-12-VBDU-0001).

REFERENCES

Chabriac, P.A., Fabbri, A., Morel, J.C., Laurent, J.P., & Blanc-Gonnet, J. (2014). A procedure to measure the in-situ water content in rammed earth and cob. *Materials*, 7, 3002–3020.

Fabbri, A., Coussy, O., Fen-Chong, T., & Monteiro, P. (2008). Are deicing salt necessary to promote scaling in concrete ? *Journal of Engineering Mechanics*, 134, 589–598.

Fraunhofer. (n.d.). IBP / WUFI. Retrieved February 08, 2013, from http://www.wufi.de/

Gray, W. (1983). General conservation equations for multi-phase systems. *Advances in Water Resources*, 6, 130–140.

Grünewald, J. (1997). *Diffusiver und Konvektiver Stoff- und Energietransport in kapillarporösen Baustoffen*. Technische Universität Dresden, Germany.

Harris, D.J. (1999). A quantitative approach to the assessment of the environmental impact of building materials. *Building and Environment*, 34, 751–758.

Künzel, H.M. (1995). *Simultaneous heat and moisture transport in building components one—and two-dimensional calculation using simple parameters—*

Report on PhD thesis, Fraunhofer Institute of Building Physics (Vol. 1995, p. 65).

Liuzzi, S., Hall, M.R., Stefanizzi, P., & Casey, S.P. (2013). Hygrothermal behaviour and relative humidity buffering of unfired and hydrated lime-stabilised clay composites in a Mediterranean climate. *Building and Environment*, *61*, 82–92.

Luikov, A.V. (1975). Systems of differential equations of heat and mass transfer in capillary-porous bodies. *International Journal of Heat and Mass Transfer*, *18* (Pergamon Press), 1–14.

Philip, J.R., & De Vries, D.. (1957). Moisture movement in porous materials under temperature gradients. *Transactions, American Geophysical Union*, *38*, 222–232.

Sameh, S.H. (2013). Promoting earth architecture as a sustainable construction technique in Egypt. *Journal of Cleaner Production*, 1–12.

Whitaker, S. (1977). *Simultaneous Heat, Mass, and Momentum Transfer in Porous Media: A Theory of Drying* (Vol. 13, pp. 119–203).

Rammed Earth Construction – Ciancio & Beckett (Eds)
© *2015 Taylor & Francis Group, London, ISBN 978-1-138-02770-1*

Structural behavior of Cement-Stabilized Rammed Earth column under compression

D.D. Tripura
Research Scholar, Department of Civil Engineering, Indian Institute of Technology, Guwahati, India

K.D. Singh
Associate Professor, Department of Civil Engineering, Indian Institute of Technology, Guwahati, India

ABSTRACT: The paper presents a novel experimental investigation, comprising material tests and column tests, focusing on the effect of concentric axial loading and slenderness on the capacity reduction factors using Cement Stabilized Rammed Earth (CSRE) columns of square (S) cross section. The test results of columns compare quite favorably with published codal provisions. There is a reduction in strength as the height-to-thickness ratio increases from about 2 to 10. The shear failures noticed in the columns resemble the shear failures of short-height prism.

1 INTRODUCTION

Rammed earth is used for construction of walls and other building components. In this technique, the temporary formwork is filled with a 10 to 12 cm moist earth (stabilized or unstabilized) layer followed by ramming and then a new 10 to 12 cm layers are added and rammed in progressive layers. The formwork is removed and placed at a higher level until the desired height is reached. A significant number of magnificent rammed earth buildings are to be found in southern India, particularly in Bangalore.

Due to limited structural design regulations for earth buildings, rules developed for masonry construction are generally followed. At present, the most well known structural design standard for earth building has been developed in New Zealand (NZS: 4297, 4298, 4299–1998), India (IS: 13827–1998), Australia (Standards Australia 2002) and the United States (ASTM: E2392/E2392M-10 -2010). Over the past 50–60 years, structural design guidance for simple earth buildings has also been published in various parts of the world, some of them are Australia (Middleton 1987), the United States (Tibbets 2001), Germany (Minke 2000) and the United Kingdom (Walker et al. 2005).

Maniatidis and Walker (2008) studied the structural capacity of unstabilized rammed earth columns of square cross section focusing on the effect of load eccentricity and slenderness, determined the capacity reduction factors in combined axial compression, and bending. Reddy and Kumar (2011) investigated the strength and structural behavior of story-high CSRE walls under compression, assessed its ultimate crushing strength considering slenderness effects, and reported that the load carrying capacity decreases with increasing slenderness.

From the detailed literature reviews, it can be concluded that the application of masonry standard design provisions to earthen walls has never been adequately validated experimentally and there are limited studies on strength and behavior of CSRE columns. Hence, the present study has been under-taken.

The primary aim of the study is to investigate the validity of using masonry design rules for the design of cement stabilized rammed earth columns. The paper presents a novel experimental investigation, comprising material tests, prism and column tests, focusing on the effect of concentric axial loading and slenderness on the capacity reduction factors using CSRE columns of square cross sections.

2 EXPERIMENTAL PROGRAM

2.1 *Material*

Table 1 outlines the properties of soil used. The properties of soil was determined as per Indian standard codes - IS 2720 Part 4 (1995), IS 2720 Part 5 (1995) and IS 2720 Part 7 (2002). Properties of the selected soil comply with general published recommendations for rammed earth construction. Ordinary Portland cement of 43-grade was used throughout the experimental investigations.

Table 1. Properties of soil used.

Soil property	Percentage value
Grain size distribution:	
Sand	79%
Silt	13%
Clay	8%
Atterberg limits:	
Liquid limit	31.7%
Plastic limit	22.9%
Plasticity index	8.8%
Compaction characteristics:	
(a) Soil	
Optimum Moisture Content (OMC)	19%
Maximum dry density (g/cc)	1.7
(b) Soil with 10% cement	
Optimum Moisture Content (OMC)	20%
Maximum dry density (g/cc)	1.8

2.2 Equipment for production of test specimen

For production of 150 mm square prisms of 300 mm height and columns of 900 to 1500 mm heights, the following equipments were used:

1. A mild steel rammer weighing 5.6 kg with a solid handle of 25 mm diameter and 1020 mm length attached with a 95 mm x 95 mm mild steel ramming face was used for compaction.
2. A wooden mould of 150 mm square section (inner dimension) with 1500 mm height having 20 mm wall thickness was fabricated and fastened with nuts and bolts and further provided with a wooden base plate for fixing the mould in position.
3. A 0.5 mm thick wall mild steel collar of 97 mm × 97 mm cross section having 300 mm height was used to facilitate the location of the rammer in the mould whenever required.

Compaction throughout the test program was carried out with the help of a compaction machine capable of compacting with a free fall of rammer height of 300 mm approximately.

2.3 Casting of prisms and columns

Prior to production of test specimens the soil sample was sun—dried, ground and pass through 4.75 mm sieve. The soil was then dry mixed with 10% cement (by mass of dry soil) before mixing with water. OMC was occasionally determined by Rapid Moisture Meter during the entire production run. The inner walls of the mould were covered with either thin polythene or sellotape to avoid adhesion of test specimen with the mould walls. The wetted mix was then poured into a mould and compacted with a rammer from a 300 mm height of fall into 10–12 cm thick layers. Frog/dent of 10–20 mm deep was provided on every compacted layer to enhance proper bonding between the successive layers. The whole process was continued until the desire height of the column was reached.

To achieve the required density, compaction energy equivalent to standard Proctor value was adopted throughout the production run. The compaction energy or effort was calculated using the formula given in ASTM D698–12 (2012) as follows:

$$E = \frac{\left(\begin{array}{c} No. \\ of\ layers \end{array}\right)(No.\ of\ blows/layer)\left(\begin{array}{c} Weigt\ of \\ rammer, kg \end{array}\right)\left(\begin{array}{c} Heigt\ of \\ drop, cm \end{array}\right)}{Volume\ of\ mould,\ cm^3}$$

(1)

where, E = Compaction Energy, Kg.cm/cm³. Within 30–40 minutes after casting, the wooden formwork was removed and the test specimens were kept in

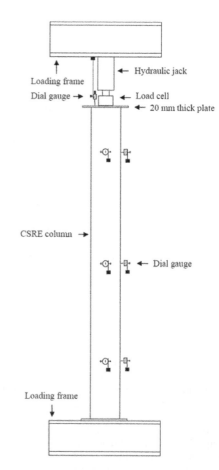

Figure 1. Column test setup.

air for 24 hours prior to wet curing. After 28 days of curing under wet burlap/gunny cloths, the test specimens were dried in air inside the laboratory for 3–4 weeks prior to testing. At least three representative samples from three different locations of the failed specimen were collected immediately in a beaker to determine the moisture content at the time of testing.

2.4 *Testing of prisms and columns*

Five prisms with a height to thickness ratio of 2 were used to determine the compressive strength. A load controlled Universal Testing Machine of 400 kN capacity was used for testing and the load was applied at a uniform rate of 2.5 kN/min up to failure.

The columns tests were comprised of three samples of three specimens with an approximate height of 900 mm (denoted S900), 1200 mm (S1200), and 1500 mm (S1500).

The columns were placed in position as shown in Figures 1 and 2. The vertical load was applied using 500 kN capacity hydraulic jack and measured using a 250 kN load cell at the rate of 2.5 kN/min until failure. Lateral movement of each column was recorded using six digital dial gauges, two

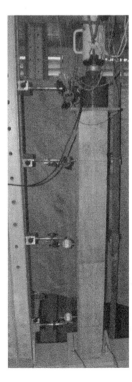

Figure 2. Column test arrangement.

at the top, two in the middle, and two at the bottom of each column placed at right angles. Furthermore, digital dial gauge was attached on top of each column, to monitor the vertical movement under incremental increasing load. All the test data were monitored and recorded both manually and automatically through a digital data acquisition system at a load interval of 10 kN continuously. As collapse was difficult to predict, some instrumentation was removed as a precaution before the ultimate load was reached.

3 RESULTS AND DISCUSSION

3.1 *Strength and failure pattern of prisms*

Test results of prisms are summarized in Table 2. The initial tangent modulus of prism is about 2 GPa. Predominantly, the prisms failed by vertical cracking followed by shearing of the material along the full height.

3.2 *Deflections and failure patterns of columns*

Although not visible by eye, lateral deflections of the columns occurred at various stages of loading (Figure 3). Maximum lateral deflections of columns were in the range of 3–4 mm for S1500 and decreases gradually to 1 mm for S1200 and 0.7 mm for S900 respectively. The failure of S1500 column was initiated by the development of vertical and inclined shear cracks at about 300–350 mm distance from top toward mid-height by spalling, splitting and shearing off portions of the column,

Table 2. Summary of test results.

Physical properties	Specimen details			
	Prism	S900	S1200	S1500
Average compressive strength (MPa)	5.3	4.07	3.72	3.65
Standard deviation (MPa)	1.29	0.54	1.30	0.46
Average moisture content at test (%)	5.01	6.43	6.23	5.55
Standard deviation (%)	0.06	0.49	0.58	1.91
Height to thickness ratio (h/d)	2	6	8	10
Slenderness ratio (l/r)	6.9	20.8	27.7	34.6
Tangent modulus at a stress level of ultimate strength (MPa)	230	627	716	730
$\sigma_{critical}$ (MPa)[a]	47.3	14.3	9.2	6.3

[a]Critical buckling stress based on tangent modulus theory.

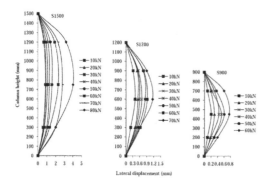

Figure 3. Lateral deflections along the height of the column at various stages of loading.

leaving a wedge shape across the thickness of the column and sometimes a vertical splitted plane. It was further observed that the shear failure occurred at about 250–300 mm height from bottom in some of the column. The columns did not show visible buckling and the shear failure patterns indicate that the column collapse was dictated by material failure. Furthermore, the deflection patterns of columns were dominated by slenderness effect, i.e., greater the slenderness ratio greater is the deflection.

3.3 Compressive strength and design of columns

Three slenderness ratios (effective height to thickness ratios) of 6, 8 and 10 with zero eccentricity were investigated. Table 2 shows the detail test results. As the slenderness ratio of the columns were increased from 6–10, the compressive strength declined by about 8.6–10% and the load carrying capacity also decreases. The stress reduction factor for a slenderness ratio of 8 and 10 are 0.91 and 0.90 respectively. These values were determined by dividing the compressive strength of respective column by the compressive strength of column having slenderness ratio 6.

Based on tangent modulus theory the ultimate strength of the CSRE columns was also estimated. Bleich (1952), Sahlin (1971), and Reddy and Kumar (2011) used the theory in their extensive studies. The buckling strength was calculated using the following formula:

$$\sigma_{cr} = \frac{P_{cr}}{A} = \frac{\pi^2 E}{(l/r)^2} \qquad (2)$$

where, P_{cr} = buckling load (N); A = cross-sectional area of the wall (mm²); E = tangent modulus at failure (N/mm²); l/r = slenderness ratio; l = effective height of the column (mm); r = radius of gyra-

tion = $\sqrt{(I/A)}$ (mm); and I = moment of inertia (mm⁴).

The E value was estimated from the stress-strain curve of prism corresponding to ultimate strength (average compressive strength) of the columns (Table 2).

The effect of slenderness ratio on the compressive strength of CSRE column is shown in Figure 4. The experimental value and the value predicted by tangent modulus theory tend to converge with each other as the slenderness ratio increases beyond 12. The difference between experimental and theoretical values for lower slenderness ratio values can be attributed to the brittle nature of failure and the absence of buckling in short columns.

3.4 Comparison of experimental and published results

The experimental capacity reduction factors under concentric loading derived from the experimental analysis are compared with published values for structural masonry as shown in Table 3. Stress reduction factor from the literature, which are based on reduction factors for masonry, are different from the actual value obtained from the present

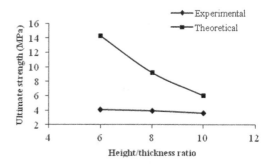

Figure 4. Effect of slenderness ratio on the compressive strength specimens.

Table 3. Comparison of experimental and published capacity reduction factors.

Column series	S900	S1200	S1500
Slenderness ratio	6	8	10
	Stress reduction factors		
Experimental	1	0.91	0.90
NZS 4297-1998	1	0.94	0.88
IS 1905-2002	1	0.95	0.89
AS 2000	1	0.94	0.88
BS 5628-1:1992	1	1	0.97
Maniatidis and Walker, 2008	1	1.41	1.24

investigation. Hence, it is important to consider the actual values of the CSRE columns or walls while designing the rammed earth structures.

3.5 *Conclusions*

Results from square column tests are presented, and the effects of column slenderness have been investigated. The following conclusions have been made:

1. The slight variation between experimental and published results of column reduction factors may be due to material properties. As rammed earth is a monolithic material therefore, its behavior under compressive loads is different from that of masonry.
2. The reduction in compressive strength and stiffness of columns is affected by variation in slenderness ratio.
3. Due to limited study it is not possible to propose the use of stress reduction factors for designing of columns, until further research is carried out.

REFERENCES

ASTM. 2010. Standard guide for design of earthen wall building systems. E2392/E2392M—10, West Conshohocken, PA.

Bleich, F. 1952. *Buckling strength of metal structures*, McGraw-Hill, New York.

Indian Standard. 1995. Determination of liquid and plastic limit. IS 2720 (Part 5), New Delhi, India.

Indian Standard. 1995. Specification for methods of test for soils-grain size analysis. IS 2720 (Part 4), New Delhi, India.

Indian Standard. 1998. Improving earthquake resistance of earthen buildings—guidelines. IS 13827, New Delhi, India.

Indian Standard. 2002. Determination of water content-dry density relation using light compaction. IS 2720 (Part 7), New Delhi, India.

Keable, J. 1996. *Rammed earth structures. A code of practice*, Intermediate Technology Publications, London, U.K.

Maniatidis, V. & Walker, P. 2008. Structural capacity of rammed earth in compression. *J. Mater. Civ. Eng.*, 20(3), 230–238.

Middleton, G. F. 1987. [Revised by Schneider, L.M. (1992)], *Bulletin 5. Earth wall construction*, 4th Ed., CSIRO Division of Building, Construction and Engineering, North Ryde, Australia.

Minke, G. 2000. Earth construction handbook. The building material earth in morden architecture, WIT Press, Southampton, U.K.

New Zealand Standard (NZS). 1998. Engineering design of earth buildings. *NZS No. 4297*, Wellington, New Zealand.

New Zealand Standard (NZS). 1998. Materials and workmanship for earth buildings. *NZS No. 4298*, Wellington, New Zealand.

Standards Australia. 2002. *Australian earth building handbook*, Sydney, Australia.

Sahlin, S. (1971). *Structural masonry*. Prentice Hall, Upper Saddle River, NJ.

Tibbets, J.M. 2001. *Emphasis on rammed earth—The rational*. Interaméricas Adobe Builder, 9, 4–33.

Venkatarama Reddy, B.V. & Prasanna Kumar, P. 2011. Structural behavior of story-high cement-stabilized rammed earth wall under compression. *J. Mater. Civ. Eng.*, 23 (3), 240–247.

Walker, P., Keable, R., Martin, J. & Maniatidis, V. 2005. *Rammed earth design and construction guidelines*, BRE Press, Bracknell, UK.

Rammed Earth Construction – Ciancio & Beckett (Eds)
© 2015 Taylor & Francis Group, London, ISBN 978-1-138-02770-1

Specimen slenderness effect on compressive strength of Cement Stabilised Rammed Earth

B.V. Venkatarama Reddy, V. Suresh & K.S. Nanjunda Rao
Department of Civil Engineering, Indian Institute of Science, Bangalore, India

ABSTRACT: Standard code procedures do not exist for the determination of characteristic compressive strength of Cement Stabilised Rammed Earth (CSRE). Correlation between specimen's slenderness (height to thickness ratio in the range of 2–6) and compressive strength for CSRE using 7% cement and having a density of 1800 kg/m³ is discussed. The results show that there is hardly any variation in the compressive strength of CSRE when height to thickness ratio of specimen is in the range of 2–6.

1 INTRODUCTION

Rammed earth structural elements are monolithic and built by compacting processed soil in progressive layers in a temporary formwork. There are two types of rammed earth structures: stabilised rammed earth and un-stabilised rammed earth. Apart from soil and aggregates the stabilised rammed earth elements contain inorganic additives such as cement or lime. Use of cement as a stabiliser for rammed earth walls has been demonstrated in many parts of the world since the last five to six decades. Examples of successful application of cement stabilised rammed earth buildings can be seen in Australia, USA, Europe, Asia and many other countries (Verma & Mehra 1950, Easton 1982, Houben & Guillaud 2003, Walker et al. 2005).

Quality of any construction material is generally assessed by measuring the compressive strength of the material. Normalised compressive strength value becomes essential for assessing the load carrying capacity of rammed earth wall or any other rammed earth structural element. There are standardised code procedures for assessing the compressive strength of commonly used conventional materials and assemblies such as bricks, blocks, masonry, etc. Standard code procedures are rare for the determination of characteristic compressive strength of CSRE. Apart from specimen size and its slenderness the compressive strength of CSRE is controlled by the cement content, density and moisture content of the specimen for a given soil composition and grading. The present study attempts to examine if there is any correlation between specimens slenderness and compressive strength for CSRE.

2 MATERIALS USED IN EXPERIMENTS

CSRE specimens were cast using Ordinary Portland Cement (OPC) conforming to IS 12269 (1987). 28 day compressive strength of OPC tested following the procedure outlined in IS 4031 (1988) was 69.2 MPa. The initial and final setting time for the cement was 148 and 312 minutes respectively.

A local red soil and sand were used in preparing the rammed earth specimens. Comprehensive investigations of Venkatarama Reddy and Prasanna Kumar (2011) on cement stabilised rammed earth revealed that the optimum clay content in the soil yielding maximum strength is about 15%. The local red soil has clay content of 43% and therefore the local red soil was reconstituted by mixing the soil and sand in the proportion of 1:2 (soil: sand, by mass). Figure 1 shows the grain size distribution

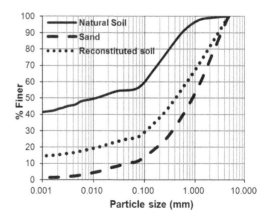

Figure 1. Grain size distribution of soil.

Table 1. Characteristics of reconstituted soil.

Soil property	Details
Textural composition (% by mass)	
Sand (4.75–0.075 mm)	73.7
Silt (0.075–0.002 mm)	11.0
Clay (<0.002 mm)	15.3
Atterberg Limits	
Liquid limit (%)	26.02
Plasticity Index	10.75
Compaction characteristics	
(a) Without cement	
Standard Proctor OMC (%)	10.2
Maximum dry density (kg/m³)	2039.0
(b) With 7% cement	
Standard Proctor OMC (%)	10.8
Maximum dry density (kg/m³)	1990.0
Predominant clay mineral	Kaolinite

curves for the natural local red soil, sand and reconstituted soil. The clay content of the reconstituted soil is 15.3%. The characteristics of the reconstituted soil are given in Table 1. Reconstituted soil was used for casting the rammed earth specimens. The soil contains kaolinite clay mineral. Liquid limit and plasticity index values for the reconstituted soil are 26.02% and 10.75 respectively. There is hardly any difference between standard Proctor Optimum Moisture Content (OMC) and maximum dry density values for the soil with and without cement.

3 METHODOLOGY AND EXPERIMENTAL PROGRAMME

The investigations involved determining the compressive strength of CSRE specimens with height to thickness ratio ranging from 2 to 6 in a displacement controlled test rig. Table 2 gives the details of the CSRE specimens. Height to thickness ratio of 2 is for cylindrical specimen. The height of CSRE wallettes having 600 × 155 mm (length × thickness) cross sectional dimensions was varied from 400–900 mm, thus there were three height to thickness ratios (2.58, 3.87, and 5.81) for the wallette specimens.

A metal mould was used for casting the wallettes and cylindrical specimen. Reconstituted soil and 7% cement (by mass) were used. Moulding moisture content was based on standard Proctor OMC. The mass of the partially saturated soil-cement mixture going into each compacted layer (of 100 mm thickness) was monitored such that the dry density of the specimen is maintained at 1800 kg/m³. The specimens were cured under wet burlap. After 28 days curing the specimens were air dried and then dried in a drying chamber at 50–55 °C till constant weight was obtained. The dried specimens were tested for compressive strength. Figure 2 shows the test set-up for testing a CSRE wallette. The wallettes were tested in a displacement controlled test rig. The moisture content of the dry specimens after the test was monitored.

Table 2. Dimensions of CSRE specimens and compressive strength.

Specimen type & size (mm) (L × T × H)	H/T ratio	M.C. (%)	σ_c (MPa)	k	σ_c^1 (MPa)
150 × 300 (cylinder)	2.00	1.50	4.61	0.00	4.61
600 × 155 × 400	2.58	2.31	4.45	0.15	4.60
600 × 155 × 600	3.87	2.44	4.40	0.18	4.58
600 × 155 × 900	5.81	1.46	4.55	−0.008	4.54

M.C.—Moisture content; σ_c—Dry compressive strength; σ_c^1—Corrected compressive strength = $k + \sigma_c$.

Figure 2. Test set-up.

4 RESULTS AND DISCUSSION

The results of the compressive strength tests on CSRE cylinders and wallettes are given in Table 2. The Table gives specimen size, height to thickness ratio, moisture content during testing, compressive strength, correction factors and corrected compressive strength. Method of obtaining the correction factor and corrected strength has been explained in the following paragraphs. The results given in the Table represent the mean of three specimens. The discussion on these results is given below.

Figure 3 shows a relationship between dry and saturated compressive strength for CSRE cylindrical specimen. Here, the cylindrical specimens were cured for 28 days, dried in air for 14 days inside the laboratory. The air dried specimens were then dried in an oven at 50 °C till constant weight was obtained. Oven dried specimens were cooled to ambient room temperature (~28 °C) and then tested for dry compressive strength. The moisture content of the dry specimens was 1.5%. The oven dried specimens were tested for strength in saturated condition by soaking them in water for 48 hours prior to testing. The saturated moisture content was 13.92%. The linear relationship shown in Figure 3 was used to get the corrected compressive strength of wallettes. The wallettes were dried in a chamber at 50 °C and allowed to cool down at ambient temperature for two days. During this period they have picked up some moisture from the air and hence the moisture content of the wallettes varies in a range of 1.46–2.44%. The compressive strength of stabilized rammed earth is sensitive to moisture content of the specimen at the time of test. In order to account for this the compressive strength of rammed earth specimens given in Table 2 were corrected with reference to cylinder strength having moisture content of 1.5%. The effect of moisture content on strength was accounted using the slope of the linear relationship shown in Figure 3 for cylinder strength. For example the correction factor for the wallette (of size: 600 × 155 × 600 mm) = (2.44 − 1.5) × (0.19) = 0.18. Here, 0.19 is the slope of the line shown in Figure 3. Therefore the corrected compressive strength of the wallette = (4.40) + (0.18) = 4.58 MPa. The correction factors for the other wallette specimens are given in Table 2.

The Height to Thickness (H/T) ratio of CSRE specimens varies between 2 and 5.81. A plot of H/T ratio versus corrected compressive strength is shown in Figure 4. The plot shows a marginal drop (<2%) in compressive strength of CSRE when H/T ratio was varied between 2 and 6. Based on this data it can be assumed that compressive strength of CSRE cylinder or prism with H/T ratio of 2 can be used for assessing the characteristic compressive strength. The investigations of Venkatarama Reddy and Prasanna Kumar (2011) show marginal variation in wallette strength (H/T = 5) when compared to prism strength (H/T = 2). However, they did not properly correct the strength values to account for small variations in moisture content of specimens.

4.1 Failure patterns of CSRE specimens

Figure 5 shows typical failure patterns for the CSRE cylinder and wallette specimens. The cylinder and wallette specimens show typical shear failure.

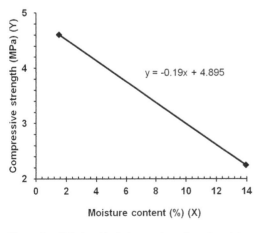

Figure 3. Relationship between strength and moisture content.

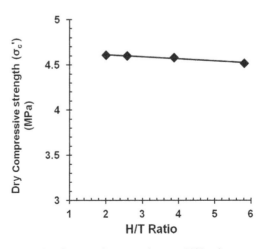

Figure 4. Compressive strength versus H/T ratio.

Figure 5.   Typical failure patterns.

5   CONCLUDING REMARKS

Influence of specimens H/T ratio on compressive strength of CSRE was discussed considering 7% cement and 1800 kg/m³ density. The results show that there is hardly any variation in the compressive strength of CSRE specimen for H/T ratio in the range of 2–6. Prism or cylinder compressive strength (H/T = 2) can be used to assess the characteristic compressive strength of CSRE. However, more test results need to be generated considering other cement contents and densities for generalising the observations made in this study.

ACKNOWLEDGEMENTS

The authors wish to thank the IFCPAR/CEFIPRA for providing funds under the Indo-French collaborative project on "Developing design guidance for rammed earth construction".

REFERENCES

Easton, David. (1st Ed.) (1982). *The rammed earth experience*. Wilseyville, CA: Blue Mountain Press.

Hall, M., (2002) *Rammed earth: traditional methods, modern techniques, sustainable future*. Build Engineer 77(11): 22–24.

Houben, H. & Guillaud, H., (2004). *Earth construction—A comprehensive guide*, London: Intermediate Technology Publications.

IS 12269 (1987), *Specification for 53 grade ordinary Portland cement*. Bureau of Indian standards, New Delhi, India.

IS 4031 (part 7) & (part 5) 1988, *Methods of physical tests for hydraulic cement*. Bureau of Indian standards, New Delhi, India.

Venkatarama Reddy, B.V. & Prasanna Kumar, P., (2011). Cement stabilised rammed earth—Part B: Compressive strength and elastic properties, *Materials and Structures*, 44(3): 695–707.

Verma, P.L. & Mehra, S.R., (1950). Use of soil-cement in house construction in the Punjab. *Indian Concrete Journal*., 24: 91–96.

Walker, P., Keable, R., Martin, J. & Maniatidis, V., (2005). *Rammed earth design and construction guidelines*, BRE Bookshop UK.

Rammed Earth Construction – Ciancio & Beckett (Eds)
© 2015 Taylor & Francis Group, London, ISBN 978-1-138-02770-1

Earth construction: Poured earth mix design

J.A. Williamson
Martin/Martin Inc., Denver, Colorado, USA

F.R. Rutz
University of Colorado Denver, Denver, Colorado, USA

ABSTRACT: Poured earth has good potential for mechanization using conventional concrete construction equipment. A mix design method, based on a modified version of the Fuller's Formula, was tested. The particle size distribution of several varieties of crusher fines were mathematically combined with the particle size distribution of a local pit mine clay. These mathematical mixes were evaluated against the modified Fuller's Formula. The mix showing the closest compliance was chosen and three test mixes were produced at varying clay content. Water content was experimentally determined for each mix that targeted a 6 inch (152.4 mm) slump. Each mix was then evaluated for shrinkage, weather resistance, compressive strength, and modulus of rupture. The mathematical mixing model was also verified by gradation analysis. Results of the test mix are summarized. Comments on poured earth mix design are offered.

1 INTRODUCTION TO POURED EARTH

Major hurdles that earth construction faces are labor cost and the variation of soil from site to site. This variation can lead to expensive case-by-case testing and evaluation of soils before they can be used.

Poured earth can be thought of as concrete with clay as the binding agent instead of cement. The proper soil mix can be placed with current concrete construction equipment and the labor and cost of earth construction can thus be minimized. The uniformity of earth as a construction material can be increased if the soils are manufactured. Recycled concrete, crusher fines, and the remainder from gravel sieving, are all potential materials to combine into standard mixes.

2 OBJECTIVE OF THIS STUDY

The purpose of this study is to investigate the design of poured earth mixes using commercially available raw or by-product materials that can be mixed and placed by conventional concrete construction equipment. The basic materials proposed for this earth mix are crusher fines for aggregate and color, and clay from quarries, mines, or deep foundation construction.

In the right proportions these elements could make a well-graded, pourable, and durable mix that meets the requirements for adobe construction of 300 psi (2068 kPa) compression strength and 50 psi

(345 kPa) modulus of rupture (New Mexico Building Code, 2009, IBC, 2006). Shrinkage and erosion resistance of this mix should meet or exceed the limits set forth in NZS 4298 (New Zealand Standards Committee, 1998).

3 DESIGN OF SOIL MIX

The mix designs were based on particle size distribution. A particle-size distribution curve that approximates the shape of the Modified Fuller Formula represented in Equation 1, will optimize particle size distribution (Maniatidis & Walker 2003).

$$a_{10} = 100\left(\frac{d}{D}\right)^n + 10 \qquad (1)$$

where a_{10} = the percent of soil mass smaller than a given diameter; d = diameter of a given grain; D = largest grain diameter in the soil sample; n = the grading coefficient, which is related to the shape of the grain shape.

Maniatidis & Walker (2003) claim most soil particles have a grading coefficient of 0.20 to 0.25. Since gradation analysis is not a very repeatable test, the values for "n" suggested by Maniatidis and Walker (2003) are believed by the authors to create too narrow of a distribution envelope. The minimum and maximum values used in this study for "n" are 0.15 and 0.3 respectively (See Figure 1).

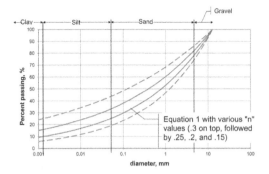

Figure 1. Theoretical mix design envelope.

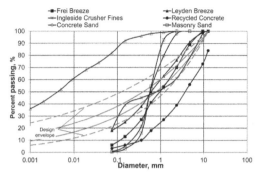

Figure 3. Aggregates used for mixes.

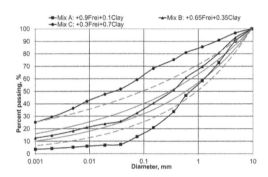

Figure 2. Test mixes.

Figure 2 shows the particle-size distribution curve for the various mix ingredients investigated for this study. The names of the clay, breeze/crusher fines indicate their origin from various pits and mines around the state of Colorado, USA. (Crusher fines and breeze are equivalent terms. They are the remaining material that is too small to use as gravel after crushing rock).

None of these materials, besides Leyden Clay, display any clay like qualities, in other words, they are non-plastic.

Leyden Clay is the binding agent in this mix and comes from a pit near a Leyden, Colorado. The Atterberg limits, which measure the plasticity of fine-grained soils, can be used in conjunction with Cassagrande's plasticity chart to determine the probable predominant clay minerals in a soil. The Atterberg Limits for Leyden Clay are a Liquid Limit of 38 and a Plasticity Index of 21. These numbers chart the clay near the U-line on Cassagrande's chart, which means the clay is probably active and mostly Montmorillonite. Therefore, a soil mix made with this clay instead of more stable clay will have more potential for drying shrinkage, but a higher compressive strength when dry.

To design a theoretical "best" mix, the various ingredients are combined mathematically by Equation 2:

$$PP_{sieve\ i}^{mix} = \Sigma\left(PP_{sieve\ i}^{ingredient\ n} \times C_{ingredient\ n}\right) \quad (2)$$

where $PP_{sieve\ i}^{mix}$ = the percent passing the ith sieve of the mix; $PP_{sieve\ i}^{ingredient\ n}$ = the percent passing the ith sieve of the nth ingredient and $C_{ingredient\ n}$ = the percentage the nth ingredient makes up of the mix. Various resulting particle size distribution curves, grouped by major aggregate ingredient, were compared visually to the design envelope.

The Frei Breeze mixes were chosen to be studied more closely. The three mixes in Figure 3 were used to make test specimens. Mix A was 90% Frei Breeze and 10% Leyden Clay by weight. Mix B was 65% Frei Breeze and 35% Clay by weight. Mix C was 30% Frei Breeze and 70% Clay by weight.

The controlling criterion for water content was slump of the mix. The target slump was 5 to 6 inches (127 to 152 mm) because that would provide an easily worked mix. Small batches of each mix were made and the water content was increased until the target slump was achieved.

For Mix A, the slump of 4.25 inches (108 mm) is deemed appropriate with a water to aggregate ratio (W/A) of 15%. The slump of 5.75 inches (146 mm) for Mix B is right on target with a W/A of 20%. Since the slump for Mix C was zero at a water content of 24% of aggregate weight and a slump of 7¼ inches (184 mm) was measured at a water content of 28%, the design mix was to be mixed at a water content of 26%.

4 TESTING DESCRIPTION

Three types of specimens were used for each mix to accomplish the required tests; blocks, cylinders, and shrinkage boxes. Seven blocks for each mix design were prepared measuring were 5.5 by 5.5

by 14 inches (140 by 140 by 356 mm). Six of these blocks were for 3-point modulus of rupture tests and the last one was for an erosion test and wet/dry appraisal test. Six cylinders for each mix design were made measuring 6 inches (152 mm) in diameter and were 12 inches (305 mm) tall. All six cylinders were for compression tests. One shrinkage box was required for each mix. The shrinkage boxes were 2 by 2 by 24 inches (51 by 51 by 610 mm).

After the rest of the tests were conducted, a gradation analysis was performed on a sample of soil one block of each mix.

As a general note, Mix A was the easiest to work with because it acted very similarly to concrete. Mix C was the hardest to work with because it was sticky and stiff. Mix B was not as sticky as C, but was still not as easy to work with as Mix A.

5   TEST RESULTS SUMMARY

5.1   Shrinkage test

The measurements in Figure 4 were taken 5 months after mixing.

5.2   Erosion test

Table 1 reports the results from the erosion test.

5.3   Wet/Dry appraisal

During each successive wetting cycle, each specimen lost comparatively less soil. Mix A lost the least soil and Mix C lost the most.

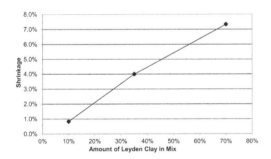

Figure 4.   Shrinkage test results.

Table 1.   Results from the erosion test.

Mix	Pit depth [mm]	Wetted depth [mm]	Erodibility
A	5	40 < 120 ok	3
B	10	25 < 120 ok	4
C	12	15 < 120 ok	4

Table 2.   Results from the compression test [kPa].

Mix	Maximum	Minimum	Average	Standard deviation
A	2234	1903	2089	113
B	2985	2799	2875	71
C	3861	3503	3689	126

Table 3.   Modulus of rupture test.

Mix	Maximum	Minimum	Average	Standard deviation
A	876	772	841	38
B	1124	1000	1062	50
C	1276	1027	1158	84

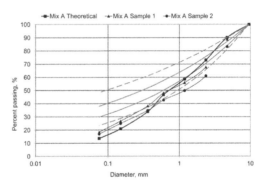

Figure 5.   Gradation analysis of mix A.

5.4   Compression test

Table 2 reports the results from the erosion test.

5.5   Modulus of rupture test

Table 3 shows the results from the modulus of rupture test.

5.6   Gradation analysis

The gradation analysis showed that the mathematically modeled mix corresponded to the physical mix. Figure 5 shows the mix A gradation results. The other two mixes showed just as much correlation.

6   TEST RESULTS DISCUSSION

The results of the shrinkage test show that only Mix A has shrinkage even close to the shrinkage standards set by NZS 4298:1998.

The erosion test showed that the lower the clay content in the soil mix, the more resistant to erosion the mix is. From observations during the tests, this may be due to the larger particles of the breeze acting as rip-rap in a riverbank. The water drops could not pick up and move the large particles of rock readily, but the fine particles of clay, once thoroughly wetted, seemed to be easily displaced. However, the permeability of the soil decreased with an increase in clay content. This is to be expected because clay has low permeability. The crazing cracks exhibited on the Mix C specimen indicate that this mix would be unsuitable for exterior construction.

Similar to the erosion test, the wet/dry appraisal demonstrated that the lower the clay content, the more resistant the mix was to damage caused by water. According to a strict interpretation of NZS 4298:1998, none of the mixes were appropriate for use because all mixes showed at least some loss of soil layers (New Zealand Standards Committee, 1998).

Taken together, the erosion test and the wet/dry appraisal show that these poured earth mixes are susceptible to weathering. However, these tests do not account for surface coatings. With an appropriate surface coating, such as earth or lime plaster, and good detailing of eaves and wall base, we believe Mix A would perform acceptably as an exterior wall.

The compressive strength and modulus of rupture of all three mixes meet or exceed the minimum requirements of the New Mexico Building Code (2009). Both of these measures of strength were found to increase as the clay content of the soil increased. This increase in strength in proportion to increase in clay content is analogous to more cement in a concrete mix.

The gradation analysis results show that Equation 2 does a good job of approximating the gradation of a design mix. Both of the gradation analysis specimens trend with the theoretical gradation and are well within the expected repeatability of a gradation analysis test.

Mix A, as computed by Equation 2, most closely matches the Modified Fullers Parabola with an *n* of 0.15. This suggests that for the shape of particles found in these mixes, 0.15 is the appropriate exponent to use for mix design purposes using the Modified Fullers Parabola. However, Mix A is the only mix that is a fully coarse-grained soil according to the Unified Soil Classification System (Das,

2002). Mix B is on the cusp of being classified as a fine grained soil and Mix C is clearly a fine grained soil. These two mixes are more controlled by the properties of the fines than the particle size distribution. Perhaps a better design tool for these two soil mixes would be the Atterberg Limits of the soil. This should be further researched.

7 CONCLUSION

We believe that this study shows that using Equation 2, it is possible to design durable, predictable, repeatable soil mixes that could be mass produced.

Of the three mixes, Mix A is the one that most closely meets the requirements set out in the Objective of this Study. Its workability is not unlike concrete. It meets the minimum compression and modulus of rupture strengths required by the New Mexico Building Code. It has the best durability of the three mixes, as measured by the erosion test and wet/dry appraisal. However, the shrinkage would need to be reduced to meet the requirements of NZS 4298. The mix may be able to be refined to reduce the shrinkage into the acceptable range of NZS 4298 by adding properly graded larger aggregate. Alternatively, with proper detailing and construction sequencing to account for this shrinkage, we believe Mix A could be used for a construction material as is.

REFERENCES

ASTM International. 2010. "E2392/E2392M Standard Guide for Design of Earthen Wall Building Systems." ASTM International, PA.

Das, B. 2002. *Principles of geotechnical engineering*, 5th ed., Brooks/Cole, Pacific Grove, CA.

Maniatidis, V., and Walker, P. 2003. *A review of rammed earth construction*. Natural Building Technology Group, University of Bath.

Minke, G. 2006. Building with earth design and technology of a sustainable architecture, Birkhäuser, Germany.

New Mexico Building Code. 2009. "Title 14-housing and construction chapter 7 building codes general." <http://www.nmcpr.state.nm.us/nmac/_title14/T14C007.htm>. (July 17, 2011).

New Zealand Standards Committee. 1998. NZS 4298:1998 Materials and workmanship for earth buildings. Standards New Zealand, New Zealand.

Rammed Earth Construction – Ciancio & Beckett (Eds)
© 2015 Taylor & Francis Group, London, ISBN 978-1-138-02770-1

Investigating the lateral capacity of wall top fixings in rammed earth materials

L.A. Wolf & C.E. Augarde
School of Engineering and Computing Sciences, Durham University, UK

P.A. Jaquin
Land Development and Exploration, Warkworth, New Zealand

ABSTRACT: Simple truss roofs for rammed earth buildings require adequate vertical and horizontal fixing and restraint to walls, and a key issue is the form of the fixing and detail used to accomplish this. There are various sources of advice, however much lacks a basis in engineering testing. In this paper we present a study of laboratory testing of rammed earth materials cast in cubes into which reinforcing bars have been secured, a bar representing a generic wall fixing. Tests are described which involved applying a lateral load to the bar while restraining the rammed earth cube in a specially adapted loading rig. The results of the tests reveal the potential effects of material choice, bar diameter and embedment depth on pre-failure and ultimate behaviour. Differing failure modes are also observed and the paper concludes with some suggested recommendations for designing horizontal fixings in rammed earth materials based on the findings of the study.

1 INTRODUCTION

One issue associated with design in Rammed Earth (RE) is the lateral capacity of fixings, to secure a roof to the top of a wall, for instance. However, it is surprisingly hard to find advice on this matter based on recent scientific research. In the literature, some guidance is available in Keable (1996) suggesting the roof frame to a RE structure should be anchored at 900 mm centres using two strands of eight gauge minimum (3.3 mm dia.) galvanized wire secured to plates embedded 450 mm into the wall, but this seems somewhat specific. In the UK there is no British Standard for RE and the only guidance one might consider concerns fixings in (fired) masonry structures. However, the mechanical properties of masonry differ significantly from those of RE and therefore this advice is impractical.

New Zealand's national code for RE (Standards New Zealand 1998) suggests roof tie down bolts should be 12 mm diameter mild steel rods threaded at the top and anchored to an 8 mm thick mild steel plate embedded in the wall. The suggested depth of embedment varies depending on both the wind speed and the weight of the roof, The loading criterion here appears to be uplift from wind and not pure lateral loading, and indeed it is likely that in many cases, the former is

more significant than any lateral effects. However it is an indication that lateral loads alone have not been considered.

Another suggestion from the literature (Walker et al. 2005) is that embedded bolts attaching roofs should be buried to a depth of 600 mm. It is suggested that ties should be inserted at least 150 mm into the wall with a clearance of 150 mm from the edge. Lindsay (2012) provides similar advice in an Australian context. Houben & Guillard's (1994) famous book also contains guidance on fixings but lacks specifics in terms of dimensions. Interestingly, they advise anchoring roofs using continuous ties as opposed to using isolated supports. In many cases it appears that guidance derives from personal experience and tradition (as is the case with much structural design advice using RE) and therefore probably includes large (& hidden) safety factors which result in potentially uneconomical solutions.

In this paper we report selected findings from a recent programme of laboratory testing completed at Durham University aiming to gain an insight into the lateral fixing problem. A simple model of a fixing in RE, comprising a single reinforcing bar ("rebar") cast into a RE cube, is used and results are presented showing changes in mechanical behaviour as variables associated with the fixing and the RE mix are varied.

2 EXPERIMENTAL METHOD

2.1 *Experimental planning*

Before testing commenced, the significant variables that would affect the lateral capacity of fixings in RE were identified. These were then categorised into two distinct groups: those associated with the fixing and those with the RE. There is mechanical similarity between the lateral fixing problem and the behavior of piled foundations under lateral loads and this motivated the choice of the following variables in this study: embedment depth of the fixing (L), diameter of the fixing (d), eccentricity of the load above the surface of the RE (e), and the cement content of the RE. For simplicity it was decided to focus on a single point fixing to the top of an RE wall, rather than trying to model a wall with multiple fixings.

2.2 *Soil & fixing properties*

An RE mixture of 30% clay and silt, 60% sand, and 10% gravel (by mass) was used in all the tests reported here. This mix can be classified as a 30*:60:10 after Smith & Augarde (2013) or 613 after Hall & Djerbib (2004). It was chosen for its high dry density and compressive strength. Initially, a water content of 9.8% was chosen, for this mix following the advice provided in Smith & Augarde (2012). However, after some initial test mixes proved unsatisfactory, a higher water content (12%) was used. The RE samples were prepared as 100 mm cubes and cement contents of 2% and 6% were used. Deformed steel rebars, with nominal diameters of 6 mm, 8 mm, and 10 mm, were used to model the fixing in the RE. This choice was made based on advice from civil engineers within the construction industry and represented a low cost fixing. 'Hilti RE 500 resin' was used to secure the rebar in place in all cases.

2.3 *Equipment*

In order that a simple bench-mounted tension testing machine could be used for the tests, a novel casing for the cubes was designed and built. Figure 1 shows a comparison of the casing designed on Solidworks to a photo of the actual casing that was manufactured. The cubes sit in the casing with the fixing horizontal so that the vertical loading arrangement in the tension tester can be used to model lateral loading of a vertical fixing in a wall.

2.4 *Testing procedure*

The following values of the dimensional variables were used in this study:

- Embedment depth, L (mm)—25, 50, 75.
- Diameter of fixing, d (mm)—6, 8, 10.
- Eccentricity of the load above the RE surface, e (mm)—25, 50, 75.

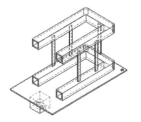

Figure 1. CAD model of the casing (left). Photograph of the manufactured casing (right).

For each cube, the RE mixture with 12% water content was produced and left in a sealed bag for 24 hours to ensure the water was evenly distributed. Cement was then added to the mixture with an additional amount of water in an attempt to maintain a free water content at compaction of 12%. The mix was then rammed using a pneumatic drill into a 100 mm steel cube mould, forming RE cubes with four layers. The cubes were immediately removed from their moulds. The procedures used to produce these RE samples followed advice on sample preparation in Hall & Djerbib (2004).

The newly constructed samples were then left for 5 days to allow the cement to cure. After this, a rebar was cut to its desired length and then the RE cube drilled and the rebar glued into place. Checks were carried out to ensure the rebar was perpendicular to the surface of the cube. The samples were then left for a further 48 hours to ensure the resin had fully hardened. The RE cubes were painted white before testing to increase the visibility of cracking during testing. Tests were then conducted using a Lloyds LR5K Plus Material Tester using displacement control at a rate of 1.0 mm/min. One cube from every batch of the soil mixture was used to obtain the compressive strength of the mix.

3 RESULTS AND ANALYSIS

Fifty-seven RE cubes were produced with 47 undergoing lateral loading tests. The average maximum compressive strength obtained for a RE mix with 6% cement content was 2.4 MPa and with 2% cement content was 1.4 MPa. These results are consistent with those from other sources.

3.1 *The effect of embedment depth*

Figure 2 shows the load vs. displacement plots for tests with varying embedment depths (one or two test results are shown for each combination). It can be seen that increasing the embedment depth of the fixing increases the lateral load the RE cube can withstand before failure, as expected. When

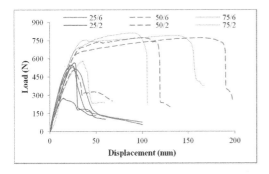

Figure 2. Results for varying embedment depth. (Legend: embedment depth/cement content).

embedment depths of 50 mm or 75 mm are used, and the RE has a cement content of 6%, the fixing bends before failure occurs (this is seen in the plots as considerable displacement before failure, without softening). This is comparable to a problem of a long piled foundation laterally loaded in soil, when a plastic hinge is formed in the pile in preference to the soil around the pile moving. Conversely at an embedment depth of 25 mm for 6% cement content, the fixing acts in a similar manner to a short pile and rotates as a rigid body, and the failure is in the RE. For a RE mixture with a cement content of 2%, all the fixings rotate comparable to a short pile despite varying embedment depths. This finding is significant as one failure is ductile (long) and the other brittle (short). Another somewhat obvious observation is that the RE samples with 6% cement content require more lateral force to displace the reinforcement than the fixings in a RE mixture with 2% cement, in line with the different material strengths. It should also be noted that the results when a cement content of 6% is used provide a clearer picture of what is happening; the results for 2% cement are not as informative as the difference in failure load is not as great. This could be due to the cement not mixing evenly across the RE mix because of the small quantity of material that is used.

3.2 Diameter of fixing

Figure 3 shows the load vs. displacement plots for RE samples with 6% and 2% cement content with varying fixing diameters, and shows that increasing the diameter of the fixing increases the lateral load at failure. Similar to the findings for embedment depth, changing the diameter of the rebar in a 6% cement content RE mix leads to changes in failure mode, i.e. for the thinnest fixing we seen ductile failure, while for the lower cement content of 2%, all failures are of the short pile type. Similarly to the previous test, RE samples with 6% cement

content require a considerably larger lateral load to displace the fixing compared to the RE samples with 2% cement content.

Both the 8 mm and 10 mm diameter fixings in the 6% cement mixture and all of the fixings in the 2% cement mixture have similar displacements before failure occurs (although with different loads). Consequently, it could be concluded that for thicker fixings when short pile failure occurs, the diameter of the fixing has limited effect on the displacement of the fixing before failure. In summary, for short pile failure mode the RE samples all fail at roughly the same fixing displacement despite varying fixing diameters.

3.3 Eccentricity of the load

Figure 4 shows the load vs. displacement plots for tests with varying load eccentricities for two different RE mixes with 6% and 2% cement content. It can be seen that increasing the eccentricity of the load decreases the lateral load the RE cube can withstand before failure occurs, as one would

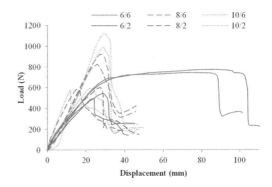

Figure 3. Results for varying fixing diameter. (Legend: fixing diameter/cement content).

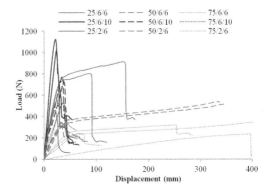

Figure 4. Results for varying eccentricity. (Legend: eccentricity/cement content/fixing diameter).

173

expect since the moment effect is greater. It should also be noted that the initial gradient of the load vs. displacement plot decreases as the eccentricity of the load increases, indicating a stiffness relation with eccentricity. When the rebar is 6 mm diameter and the RE has a cement content of 6%, a long pile failure mode is recognised for all eccentricities. At a cement content of 6% and eccentricity of 50 mm and 75 mm, the RE cubes never failed. Instead the fixing continued to bend until the test was manually aborted.

For the RE samples with 6% cement content and 10 mm diameter rebar, the failure mode is recognised as a short pile failure. For 2% cement content and a 6 mm diameter rebar, the failure mode for all but one of the results can be identified as a short pile failure mode. However, for one of the tests with a 75 mm eccentricity, the long pile failure mode was recognisable throughout loading. Once again, the results from using 2% cement are not as easily analysed compared to those with a greater cement content.

3.4 *Ratio of variables*

An analysis was carried out on the results in terms of ratios of variables as this might provide some useful guidance for real RE fixings. For RE samples with 6% cement content, it is found from this study that all samples with L/d greater than 8.3 show a long pile (ductile) failure mode. This is illustrated in Figure 5. In contrast, the ratio e/d did not appear significant as a predictor of the failure mode, however a slight tendency was noticed towards a higher e/d ratio having a long pile failure mode. The ratio L/e (embedment depth/eccentricity) ratio appears to have almost no influence on the failure mode of the fixing. For RE mixes with 2% cement content, the vast majority of the fixings had a short pile failure mode. Further dimensional analysis and ANOVA were carried out on these results but there is insufficient space to report them here.

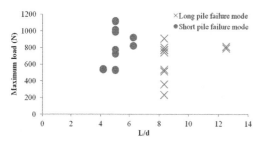

Figure 5. Maximum loads against embedment/diameter ratios, 6% cement content.

4 CONCLUSIONS

This fairly rudimentary study, based on carefully conducted scientific testing, provides some initial guidance on lateral loads on fixings in RE. Two failure modes of these fixings have been observed, one similar to that of a short piled foundation and the other to a long piled foundation. During long pile failure the reinforcement bends around a plastic hinge whereas for a short pile failure mode the pile rotates as a rigid body. It also appears that the ratio L/d is the most significant measure to allow prediction of failure mode. The practical conclusion from this research is that a fixing with a larger L/d ratio should be used thus allowing the rebar to form a plastic hinge rather than rotate as a rigid body. This means that instead of the RE failing suddenly, the fixing will instead bend, thus allowing more time for collapse to occur, i.e. a ductile failure which is to be preferred in structural design. The long pile failure mode can be considered as a serviceability state limit as opposed to the short pile failure mode which can be classed as an ultimate limit state.

Clearly the next step in this work is to consider testing with a wider variety of RE mixes, to consider the effect of fixings buried within thicker (and more realistic) layers of RE and to also consider the fixing spacing along a wall as there is likely to be interaction between fixings which will affect the overall capacity.

REFERENCES

Hall, M. & Djerbib, Y. 2004. Rammed earth sample production: context, recommendations and consistency. *Construction and Building Materials* 18:281–286.

Houben, H. & Guillard, H. 1994. *Earth construction a comprehensive guide*. London: ITDG Publishing.

Jaquin, P.A. & Augarde, C.E. 2012. *Earth building: history, science and conservation*. Bracknell: IHS BRE Press.

Keable, R. 1996. *Rammed earth structures a code of practice*. Intermediate Technology Publications.

Lindsay, R. 2012. Australian modern earth construction. In Hall et al. (eds) *Modern Earth Buildings*. Cambridge: Woodhead Publishing.

Maniatidis, V. & Walker, P. 2003. *A review of rammed earth construction*. University of Bath, Bath.

Smith, J.C. & Augarde, C.E. 2013. A new classification for soil mixtures with application to earthen construction. ECS Technical Report 2013/04 (www.dur.ac.uk/resources/ecs/research/technical_reports/SMCTechnicalPaper.pdf).

Smith, J.C. & Augarde, C.E. 2013. Optimum water content tests for earthen construction materials. *ICE Proceedings: Construction Materials* 167(2): 114–123.

Standards New Zealand 1998. *NZS 4297 Engineering design of earth buildings*. Wellington: Standards New Zealand.

Walker, P., Keable, R., Martin, J. & Maniatidis, V. 2005. *Rammed earth design and construction guidelines*. Watford: BRE Bookshop.

Rammed Earth Construction – Ciancio & Beckett (Eds)
© 2015 Taylor & Francis Group, London, ISBN 978-1-138-02770-1

Potential of existing whole-building simulation tools to assess hygrothermal performance of rammed earth construction

M. Woloszyn & A.-C. Grillet
LOCIE, UMR 5271, CNRS, Université Savoie Mont Blanc, Chambéry, France

L. Soudani, J.-C. Morel & A. Fabbri
LGCB-LTDS, UMR 5513, CNRS, Université de Lyon, ENTPE, France

ABSTRACT: The present paper investigates existing studies on hygrothermal performance of rammed-earth constructions. First a short overview of existing simulation tools at whole building level, able to deal with hygrothermal phenomena is presented. Then, existing cases of hygrothermal analysis of rammed earth construction are discussed. Finally, some recommendations are given for hygrothermal simulations of rammed earth constructions at whole building level. It appears clearly that additional researches are needed to analyze and quantify precisely the hygrothermal behavior of rammed earth walls, and their impact on indoor conditions. In parallel, many advanced simulation tools are available, and their use should be encouraged.

1 INTRODUCTION

Earth and timber are among the oldest materials used by the humans to construct dwellings and shelters. Through ages and regions, different types of constructions were developed, such as for example adobe, rammed earth, sometimes stabilized with lime or cement, timber frame with straw/earth filling, etc.

The diversity is very large, both at material scale, due to the very local origin, and at building scale, due to types of construction and shapes.

Published engineering research on rammed earth construction concerns mainly material properties and characterization, with a focus on mechanical, structural properties as well as on the durability. In parallel, embodied energy was also studied from life-cycle point of view, showing better performance of houses built with local material when compared to traditional, modern construction (Morel et al. 2001).

With the tightening of thermal regulations aiming at preventing climate change, a growing number of works is devoted to energy performance of rammed earth as a construction material. At material scale, Hall and Allinson (2009) presented data of thermal conductivity as a function of moisture content for different soil grading. They showed the importance of moisture content in the variations of thermal conductivity. The same authors (Hall and Allinson, 2008) have also investigated the response of stabilized rammed earth

fabric to cyclic thermal loading. They showed the importance of thermal mass, which can be used to improve thermal comfort in buildings in a passive way. They also investigated the impact of moisture content on thermal transmittance. This study was based on only theoretical calculations (no experiment) using thermal properties for fixed moisture gradients. However, no coupling of heat and mass transfer was considered, and thermal cycling taken into account was simplified, perfectly sinusoidal, such as required by the use of thermal admittance method. The impact of moisture content on thermal conductivity and effective specific heat was considered, however no impact of latent heat due to moisture movement inside the construction was taken into account. The authors concluded that further investigations of the hygrothermal behavior of stabilized earth materials are required in order to better understand how they interact with and control air temperature and relative humidity.

There is now a growing interest in the investigations of possible contributions of rammed earth building envelopes to the reduction of energy used to ensure good thermal comfort in buildings for heating, cooling and managing humidity level. High thermal mass as well as the strong coupling between heat and moisture transfers seem to be the major elements (Martin et al, 2010, Allison and Hall 2010). In order to analyze precisely this contribution, it is anticipated that dynamic hygrothermal phenomena at whole building scale should be investigated, both using experimental measurements and numerical simulations. In this paper, after an overview of whole-building simulation tools, existing works on hygrothermal performance of rammed-earth buildings are reported. Finally, some indications for future works are proposed.

2 WHOLE BUILDING SIMULATION TOOLS

Numerous energy performance simulation tools for Whole Buildings have been developed in the last decades. Most enable at least computing of instant indoor temperature and/or heating/cooling demand, under the combined dynamic effect of occupancy (internal loads), weather conditions, together with transfers through building's envelope, as well as some representation of HVAC (Heating, Ventilating and Air Conditioning) systems.

In September 2014, 417 building software tools for evaluating energy efficiency, renewable energy, and sustainability in buildings were described in http://www.eere.energy.gov/buildings/tools_directory/. However, only few of these tools are able to predict combined Heat, Air and Moisture (HAM) transfers in buildings. The development of these tools has been encouraged by the collaborative project Annex 41—MOIST-ENG ("Whole building heat, air and moisture response" of the International Energy Agency project, ECBCS) and was reported in Woloszyn and Rode (2008). Some of the tools have been further developed and successfully used since then. Those that have been most frequently used recently for whole building hygrothermal simulations are reported below. All of them include heat, air and moisture balance in the indoor air. When modeled, heat and moisture transfers in the enclosures are calculated in one dimension.

First, we should mention EnergyPlus (www.energy-plus.org) which is primarily an energy performance simulation engine with the possibility of association of graphical interfaces. This is a very powerful tool, more and more used throughout the world. Some key capabilities include configurable modular systems integrated with heat balance-based zone simulation, multiple comfort models, daylighting and advanced fenestration, multi-zone airflow, displacement ventilation, flexible system modeling, and photovoltaic and solar thermal simulation. At present, there are three possibilities to model transfers through the building envelope: transfer function (heat only), finite difference for heat only, and finite volumes for heat and moisture transfers (HAMT, available since September 2011, and used for example in Spitz et al, 2013, to model a wooden-frame construction).

ESP-r (www.esru.strath.ac.uk) is one of the oldest tools, able to deal with hygrothermal phenomena. It is capable of modeling the heat, power and fluid flows, within combined building and plant systems when subjected to control actions, as well as visual and acoustic performance of buildings. It has been successfully used for modeling moisture transfer and mold growth (Clarke et al, 1999, Clarke 2013).

IDA, PowerDomus and HAM-Tools, are also well known software in the community of hygrothermal simulations. IDA Indoor Climate and Energy (http://www.equa.se) is fundamentally a tool for simulation of building energy consumption. It covers a large range of phenomena, such as the integrated airflow network and thermal models, CO_2 and moisture calculation, vertical temperature gradients and daylight predictions. To calculate moisture transfer in IDA ICE, the common wall model should be replaced with a specific HAM model. Most hygrothermal simulations done with IDA were devoted to wooden construction in Scandinavian climate (Hameury, 2005, Kurnitski et al., 2007).

PowerDomus (www.pucpr.br/LST) solves heat and moisture transfer in walls simultaneously, according to a method developed by (Mendes and Philippi 2005). The model has an integrated simulation of HVAC systems. Several levels of calculation complexity in HAM models are possible (e.g. with or without moisture transfer; constant or variable material hygrothermal properties; vapor pressure or moisture content driving potentials). It has been used for hygrothermal simulations of buildings, but also of the ground and building foundations (Mendes et al., 2005, Dos Santos and Mendes 2006).

A library devoted to HAM simulation was developed in Matlab/Simulink environment and named HAM-Tools (www.ibpt.org, Sasic Kalagasidis et al. 2007). It has been used by a few research teams to simulate hygrothermal performance of buildings, mainly lightweight constructions (Piot et al. 2009, Labat et al.).

Wufi®Plus is a recent whole-building extension of a well known Wufi® software, originally limited to envelope calculations. Wufi®Plus is able to simulate hygrothermal transfers in building envelopes, together with indoor environment and energy use in the building (Holm and Lengsfeld, 2007). Its users' friendly interface, together with the popularity of Wufi 2D tool, makes Wufi®Plus popular in consulting offices.

We should also mention here one of the oldest and most popular WB tools: TRNSYS program (TRaNsient SYstems Simulation, sel.me.wisc.edu/trnsys/). It has a modular structure. The TRNSYS library includes many of the components commonly found in thermal and electrical energy systems, such as solar systems, low energy buildings, HVAC systems, renewable energy systems, cogeneration, fuel cells, etc. It also allows for predictions of the indoor relative humidity, including some buffering effect of materials, using the penetration depth model. Some examples of specific types, dedicated to HAM transfers in envelopes (e.g. Steeman et al. 2010), have been published in the literature, however none of them is available in the standard libraries, nor was used for rammed-earth walls.

Many validation and case studies are described in the literature using theses simulation tools. In general, correct performance regarding temperature and energy calculations is reported, as well as correct estimations of indoor air relative humidity (Woloszyn et al, 2009). However, only a very limited number of studies compare hygrothermal values both at room and wall level (see for example Labat et al, 2013). Moreover, for highly hygroscopic materials and dynamic boundary conditions, more discrepancy is reported for humidity calculations.

3 ENERGY PERFORMANCE OF RAMMED EARTH CONSTRUCTION

Energy used for ensuring good hygrothermal comfort was of interest in the present study. The first result is the limited number of papers devoted to rammed earth construction, in the flourishing field of building energy performance papers. The most important are described below.

Taylor et al. (2008) investigated an office building in Australia, with stabilized-rammed earth walls. The authors used experimental investigations in an occupied building, and TRNSYS modeling to deepen the analysis. In this work, investigations focused mainly on HVAC system, and very little attention was devoted to rammed earth envelope.

Ip and Miller (2011) describe a very specific example, the 'Earthship' in UK, constructed largely from recycled and reclaimed materials. In this building rammed earth is used for massive walls (some of them include recycled tyres) and in a thermal store, which acts as a seasonal heat buffer to regulate the room temperature. Initial findings have demonstrated the effectiveness of the thermal charging and discharging of the rammed earth thermal mass, which appeared to moderate the extreme external temperatures. The importance of good balance between thermal mass and solar heat gains was stressed by the authors, who suggested the use of computer simulations. This was done by Freney et al. (2013), who used Energy-Plus to model thermal performance of Earthship

building. The model was successfully calibrated using monitored results from New Mexico. Only heat transfers (no moisture) through the envelope were modeled. Excellent energy performance was shown.

An extensive study using numerical simulation was conducted by Parra-Saldivar and Batty (2006), using whole building heat simulation tool, TAS (Thermal Analysis System). This software assumes only one-dimensional heat transfer through the wall elements. An attempt to model the effects of the variation of moisture contents throughout the wall was done by dividing the wall model into three layers with different thermal conductivities. The indoor environmental performance and the energy consumption of an adobe building was assessed for three different latitudes in Mexico. The results showed the importance of modeling of massive internal walls on the attenuation of the indoor temperature fluctuations. Moreover, during the cold part of the year, the external wall thermal conductivity was the most significant variable.

Allison and Hall (2010) analyzed the hygrothermal behavior of Stabilized Rammed Earth (SRE) small building in the UK. It is probably the first paper where a whole building HAM simulation tool (WUFI®Plus) is used to investigate the behavior of a rammed earth construction. The validity of the simulation was checked using experimental measurements of temperature and relative humidity of the indoor air. The parametric study showed high moisture buffering potential of SRE walls. Some complementary works are now needed, in order to extend the validation of the model to the transfers within the walls.

4 DISCUSSION AND RECOMMENDATION

This brief overview of existing works shows, on one hand, the need of precise assessment of energy performance of rammed-earth construction, and on the other hand, numerous simulation tools, able to deal with hygrothermal transfers in building envelopes. A very important question is now "Are the existing tools adapted for the simulations of rammed earth construction?". Below, some elements, that appear important, are discussed, based on literature review and on the previous experience of the co-authors.

Modeling of massive walls: as the high thermal mass is an advantage of rammed earth construction (Collet et al. 2006, Martin et al. 2010), this point must be represented by the tool. As it can be assessed by dynamic simulation, all of the above mentioned tools are able to take this effect into account. However, in some cases of very massive walls, the use of transfer function should be done with care (Li et al. 2009).

Coupled heat and mass transfers for the assessment of moisture impact on thermal properties and moisture buffering capacity of the wall can be modeled by hygrothermal tools. However this modeling is not always straightforward. First, detailed material properties are needed (vapor permeability, liquid conductivity, sorption isotherms...), that can be costly and difficult to measure. Second, existing models still need additional experimental validations in the case of rammed earth walls. There is still no consensus on some of the modeling hypotheses (Soudani et al, 2014a and 2014b) such as the use of latent heat of evaporation instead of the latent heat of sorption, modeling of hysteresis in sorption isotherm, etc. As shown in some works concerning wood, these phenomena might be important, and therefore further investigations are needed.

Raising damp, as well as *built-in moisture,* require modeling of liquid transport, as well as modeling of at least two-dimensional transfers: the vertical movement of raising damp, and horizontal movement of heat and mass due to weather and indoor loads. None of the above mentioned tools can perform such complex simulations at whole building level. An interesting way of analyzing it is to perform a co-simulation, using two simulation tools: the first one at whole building level, with 1D HAM transfers through the envelope, and second for 2D HAM transfers in the envelope, such as proposed by Taylor et al (2013) to study flooded constructions.

Interaction between the envelope and HVAC system can be represented by any of WB HAM tools. However, published results seem to show that the impact of HVAC system is much more important than the impact of building envelope, except in the case of systems controlled by relative humidity (Taylor et al. 2008, Woloszyn et al. 2009).

5 CONCLUSIONS

Precise simulation tools are essential in order to assess the performance of rammed earth construction in terms of energy and hygrothermal comfort. Such tools are necessary to validate the compliance with energy standards as well as benefits and drawbacks in comparison with different types of building's envelope.

From analysis of published papers, it appears clearly that additional researches are needed to analyze and quantify precisely the hygrothermal behavior of rammed earth walls, and their impact on indoor conditions. Additional validations studies are necessary, combining measurements at both wall level and room level. In parallel, many advanced simulation tools are available, and their use should be encouraged. They are able to deal with the effect of thermal mass as well as with the interaction between building envelope and HVAC systems concerning heat transfers.

ACKNOWLEDGEMENTS

This work has been supported by the French National Research Agency through programs "Habitat intelligent et solaire photovoltaïque" (project HYGROBAT ANR-10-HABISOL-00X) and "Villes et Bâtiments durables" (project PRIMATERRE ANR-12-VBDU-0001-03).

REFERENCES

Allison, D., & Hall M. 2010. Hygrothermal analysis of a stabilised rammed earth test building in the UK. *Energy Build* 42: 845–52.

Clarke, J.A., et al. 1999. A technique for the prediction of the conditions leading to mould growth in buildings. *Building and Environment* 34.4: 515–521.

Clarke, J. 2013. Moisture flow modelling within the ESP-r integrated build. perf. simulation system. *J. of Build. Perf. Sim.* 6.5: 385–399.

Collet, F., Serres, L., Miriel, J., & Bart, M. 2006. Study of thermal behaviour of clay wall facing south. *Build. Env.*, *41*(3), 307–315.

dos Santos, G.H., & Mendes, N. 2006. Simultaneous heat and moisture transfer in soils combined with build. sim. *En. Build.* 38(4), 303–14.

Freney, M., Soebarto, V., & Williamson, T. 2013. Thermal comfort of global model Earthship in various European climates. In *IBPSA Conference (Chambery, France) BS2013.*

Hall M., & Allinson D. 2008. Assessing the moisture-content-dependent parameters of stabilised earth materials using the cyclic-response admittance method. *Energy and Buildings* 40: 2044–2051.

Hall M., & Allinson D. 2009. Assessing the effects of soil grading on the moisture content-dependent thermal conductivity of stabilised rammed earth materials. *Applied Thermal Engin.* 29: 740–747.

Hameury, S. 2005. Moisture buffering capacity of heavy timber structures directly exposed to an indoor climate: a numerical study. *Building and Environment,* 40(10), 1400–1412.

Holm, A., & Lengsfeld, K. 2007. Moisture-buffering effect—experimental investigations and validation. In *Proc. Buildings X conference. Clearwater Bach, FL, USA.*

Ip K., & Miller A. 2009. Thermal behaviour of an earth-sheltered autonomous building—The Brighton Earthship. *Ren. En.* 34: 2037–2043.

Kurnitski, J., Kalamees, T., Palonen, J et al. 2007. Potential effects of permeable and hygroscopic lightweight structures on thermal comf. and perceived IAQ in a cold climate. *Indoor air,* *17*(1), 37–49.

Labat, M., Woloszyn, M., Garnier, G., et al. 2013. Simulation of coupled heat, air and moisture transfers in an experimental house ... In *IBPSA Int. Conf. BS2013* Chambéry, France.

Li, X.Q., Chen, Y., Spitler, J.D., & Fisher, D. (2009). Applicability of calculation methods for conduction transfer function of building constructions. *Int. Journal of Thermal Sciences*, *48*(7), 1441–1451.

Liuzzi, S., Hall, M.R., Stefanizzi, P., & Casey, S.P. 2013. Hygrothermal behaviour and relative humidity buffering of unfired and hydrated lime-stabilised clay ... *Building and Environ.* *61*, 82–92.

Martin S, Mazarron F, & Canas I. 2010. Study of thermal environment inside rural houses of Navapalos (Spain): The advantages of reuse buildings of high thermal inertia. *Constr Build Mater* 24: 666–76.

Mendes, N., & Philippi, P.C. (2005). A method for predicting heat and moisture transfer through multilayered walls based on temperature and moisture content gradients. *Int. Journal of Heat and Mass Transfer*, *48*(1), 37–51.

Mendes, N., Oliveira, R.C.L.F., & Santos, G.H. 2005. Energy efficiency and thermal comfort analysis using the Powerdomus ... In *Proc. of the 9th Build Sim Conf (IBPSA'05)* Vol. 1, pp. 9–16.

Morel J.C., Mesbah A, Oggero M, & Walker P. 2001. Building houses with local materials: means to drastically reduce the environmental impact of construction. *Building and Environment* 36: 1119–1126.

Parra-Saldivar, M.L., & Batty, W. 2006. Thermal behaviour of adobe constructions. *Building and Environment*, 41(12), 1892–1904.

Piot A., M. Woloszyn, J. Brau, & C. Abelé. 2011. Experimental wooden frame house for the validation of whole building heat and moisture transfer numerical model. *Energy and Build*, 43, 6, 1322–1328.

Soebarto, V. (2009). Analysis of indoor performance of houses using rammed earth walls. *Building Simulation 2009*.

Soudani, L., Fabbri, A., & Woloszyn, M., et al. 2014a. Etude de la pertinence des hypothèses dans la modélisation du comportement hygrothermique du pisé. *IBPSA France*, Arras (France).

Soudani, L., Fabbri, A., Woloszyn, M., Grillet, A.-C., Chabriac, P.-A., & Morel, J.-C. 2014b. On the relevance of neglecting the mass vapor variation for modelling the hygrothermal behavior of rammed earth. *IREC Conference*.

Spitz, C., Woloszyn, M., Buhé, C., & Labat, M. 2013. Simulating combined heat and moisture transfer with EnergyPlus: an uncertainty study and ... In *IBPSA Conference (Chambery, France) BS2013*.

Steeman, M., Janssens, A., & Steeman, H.J. et al. 2010. On coupling 1D non-isothermal heat and mass transfer in porous materials with a multizone ... *Build. and Environ.* 45(4), 865–877.

Taylor, J., Biddulph, P., & Davies, M. et al. 2013. Using building simulation to model the drying of flooded building archetypes. *Journal of Building Performance Simulation*, 6(2), 119–140.

Taylor P., Fuller, R.J., & Luther, M.B. 2008. Energy use and thermal comfort in a rammed earth office building. *En. Build.* 40: 793–800.

Woloszyn M., & Rode, C. 2008. Tools for Perf. Simulation of Heat, Air and Moisture Conditions of Whole Build. *Build. Sim.* 1: 5–24.

Woloszyn M., Kalamees, T., M.O. Abadie, M. et al. 2009. The effect of combining a relative-humidity-sensitive ventilation system with the moisture-buffering capacity ... *Build. Env.* 44(3), 515–524.

Author index

For Product Safety Concerns and Information please contact our EU
representative GPSR@taylorandfrancis.com Taylor & Francis Verlag GmbH,
Kaufingerstraße 24, 80331 München, Germany

Printed and bound by CPI Group (UK) Ltd, Croydon, CR0 4YY
01/05/2025
01858470-0003